Multifunctional TiO$_2$ Nanotubes:

Doping Strategies, Structural Engineering, and Applications in Energy and Environment

Hakimeh Ghazaie, Dr. Shahab Khameneh asl,
Dr. Shahin Khameneh asl, Dr. Saeid Kakooei

ISBN: 978-1-962443-18-0

DEDICATION

To those who believed in me before I believed in myself —
and to all the quiet moments of persistence that made this work possible.

CONTENTS

ACKNOWLEDGMENTS

The author wishes to sincerely thank Dr. Shahab Khameneh Asl for his valuable guidance, constant support, and kind encouragement during the preparation of this work.

CHAPTER 1: FUNDAMENTALS OF TITANIUM AND TITANIUM DIOXIDE

1- Introduction

Titanium (Ti) is one of the most abundant transition metals in the Earth's crust, with a concentration of about 0.63% by weight. Owing to its low density, high mechanical strength, and excellent corrosion resistance, titanium and its compounds have attracted significant attention across structural, biomedical, and energy-related applications [1,2]. Among its various oxides, titanium dioxide (TiO_2) is the most widely studied due to its versatile physicochemical properties. TiO_2 exists in several polymorphic forms, including rutile, anatase, and brookite, of which rutile and anatase are the most thermodynamically stable and technologically relevant [3].

The crystallographic structure of TiO_2 plays a decisive role in determining its optical, electronic, and catalytic behavior. For example, rutile possesses a direct band gap of ~3.0 eV, whereas anatase has an indirect band gap of ~3.2 eV, making the latter more efficient in photocatalytic applications [4,5]. These fundamental differences highlight the importance of understanding both titanium as a parent metal and the structural chemistry of its oxides before exploring their nanostructures and advanced functionalities. This chapter therefore introduces the basic characteristics of titanium, the general features of TiO_2, and the crystallographic structures of its most common polymorphs (rutile and anatase), providing the foundation for subsequent chapters.

2- Titanium

Titanium (symbol Ti, atomic number 22) is a silvery-gray transition metal known for its exceptional strength-to-weight ratio and low density. One of its most valued properties is its high corrosion resistance, making it suitable for use in environments exposed to seawater, chlorine, and even aqua regia. Discovered in 1791 by William Gregor and later named by Martin Heinrich Klaproth, titanium is primarily derived from mineral ores such as rutile and ilmenite. These ores are widely distributed across the Earth's crust and lithosphere. Trace amounts of titanium also occur in rocks, soil, water, and living organisms [6].

Industrial extraction of titanium from its natural ores primarily involves the Kroll process, with the Hunter process used less frequently. The most widely used titanium compound is titanium dioxide (TiO_2), a well-established photocatalyst and a primary component in white pigment production. Other important compounds include titanium tetrachloride ($TiCl_4$), a critical catalyst in polypropylene synthesis [7,8].

"Titanium alloys, commonly incorporating elements like aluminum or vanadium, are valued for their high strength-to-weight ratio. These materials are especially relevant in aerospace and biomedical fields due to their mechanical resilience and corrosion resistance. Titanium's applications also span military, petrochemical, desalination, pulp and paper, and automotive industries. Its excellent biocompatibility has led to its widespread use in medical devices, including orthopedic and dental implants, as well as surgical instruments [7].

Even in its pure form, titanium offers strength comparable to steel at a much lower density [9]. It exhibits two allotropic forms and exists in five naturally occurring isotopes, with Ti-48 being the most abundant (around 73.8%) [10]. While titanium and zirconium share a group on the periodic table and have identical valence electron configurations, they differ considerably in chemical and physical behavior.

Titanium is primarily used in two forms: metallic titanium and titanium dioxide (TiO_2) [11]. Due to the complexities involved in its extraction and purification, the metallic form is less commonly used. In contrast, titanium dioxide sees extensive industrial application, accounting for approximately 90% of the global titanium consumption [8]. Nowadays, titanium is recognized as a strategic metal, extensively used in aircraft engines, internal aircraft structures, transportation systems, chemical industries, power generation units, alloy production, submarine construction, chemical processing plants, cooling systems, and in both nuclear and thermal power stations, among various other applications [12]. Titanium dioxide (TiO_2), by contrast, is predominantly used as a white pigment in the paint and coatings industry. In addition to this primary application, it also holds substantial importance in the production of ceramics, plastics, paper, and various electronic components [13]. Notably, TiO_2 consumption in developed countries is approximately ten times greater than in developing nations.

Titanium metal, whether in its pure form or as a low-impurity alloy, exhibits excellent resistance to corrosive environments. It is widely used in desulfurization units and oil refinery systems, in equipment associated with oil wells, and in various essential connectors. It is also applied in the medical field for manufacturing implants and surgical tools. In recent years, titanium-coated steel sheets have been produced globally due to their outstanding corrosion resistance. These coated materials have found extensive application in the oil industry, particularly in desulfurization processes within petroleum refineries. Furthermore, titanium plays a major role in the aerospace sector, especially in aircraft manufacturing. In addition to its industrial significance, titanium is used in the synthesis of titanium carbide and in advanced ceramic and electrochemical processes. It is essential in the production and recycling of metallic sheets, oil refining, gas desulfurization, seawater desalination, and water purification. Titanium components are also used in civil engineering, building construction, medical implants, dental applications, automobile manufacturing, and the development of specialized storage systems for hazardous or radioactive materials. In addition, titanium is utilized in the production of reinforced fibers for metal matrix composites, as well as in industrial-grade connectors and fasteners. It also contributes to the development of advanced high-performance alloys engineered for improved energy storage capabilities and enhanced thermal conductivity. In the jewelry industry, titanium's alloys offer desirable properties such as durability, brightness, and biocompatibility. Raw materials such as ilmenite and titanium oxides are commonly employed in the production of critical and strategic titanium-based alloys [8].

Titanium alloys are widely used in the construction of military aircraft fuselages, spacecraft, missiles, aircraft engines, and advanced weaponry, as well as in turbines, bicycles, and even laptop components. These alloys are often formed by combining titanium with elements such as aluminum, iron, manganese, molybdenum, and others [14]. Ilmenite, as a key source of titanium oxide, is further utilized in industries such as paint, paper, and plastics to provide brightness and color. It is also employed in glazing metal surfaces, glass, ceramics, fiberglass, and electroceramics [15].

Despite titanium's diverse range of applications, only around 5% of global titanium production is allocated to its metallic form. The remaining 95% is devoted to titanium dioxide, primarily as a pigment [16]. Titanium dioxide (TiO_2) primarily crystallizes as rutile or anatase. These polymorphs are distinguished by their optical and chemical stability, making them suitable for pigment production and photocatalytic applications. These properties make TiO_2 one of the most important white pigments in industries such as paint, paper, plastics, and rubber [17].

Titanium dioxide powder, particularly in its rutile form, is used to produce compounds such as titanium peroxide, titanium salicylate, and titanium tannate, which act similarly to zinc oxide on human skin. Titanium dioxide is commonly applied to soothe skin irritations and burns, and due to its ability to reflect ultraviolet (UV) radiation, it is widely used in the formulation of sunscreens and sunburn protection lotions. Additionally, TiO_2 powder is used in the manufacture of pharmaceutical capsule shells and as a coating agent for tablets to improve their appearance and stability [18].

In the cosmetics industry, titanium dioxide is widely employed in the formulation of personal care products. Globally, the annual consumption of titanium and its derivatives is estimated to range between 10^5 and 10^6 tons, with nearly 95% of this amount being used in the form of titanium dioxide (TiO_2). Renowned for its intense whiteness and excellent opacity, TiO_2 serves as a permanent pigment in products such as paints, plastics, and paper. Additionally, paints containing titanium dioxide exhibit high reflectivity in the infrared region, a property that renders them particularly valuable in applications related to astronomy [19].

Owing to its excellent strength-to-weight ratio, superior resistance to corrosion, and stability under extreme temperatures, titanium has become a material of choice in marine engineering. It is particularly valued in the production of components such as aircraft propeller shafts and marine propellers, where prolonged exposure to seawater demands both mechanical durability and chemical resilience. Furthermore, titanium and its alloys are widely used across various advanced technologies, owing to their outstanding mechanical properties, high corrosion resistance, biocompatibility, and non-toxicity. The unique combination of these properties has fueled increasing research efforts focused on the fabrication of titanium dioxide nanoparticles, which hold promise for numerous scientific and industrial applications across various fields [20].

3- Titanium Dioxide (TiO₂)

Titanium dioxide (TiO_2), also known as titanium oxide or titania, is among the most thoroughly investigated semiconductor materials, largely due to its outstanding physicochemical characteristics. In its nanoscale form, TiO_2 retains all the intrinsic properties of its bulk counterpart, while benefiting from a drastically reduced particle size. This nanoscale dimension results in a significantly higher surface area-to-volume ratio, enhancing surface reactivity and boosting overall performance across a wide range of applications [21].

TiO_2-based materials occupy a prominent position as biomaterials, photocatalysts, and key components in solar cell technologies. The exceptional functional versatility of TiO_2 primarily stems from its intrinsic bandgap characteristics and the favorable alignment of its conduction and valence band edges, which render it highly effective for a broad spectrum of redox reactions. Among the diverse morphological forms of TiO_2 nanostructures, one-dimensional (1D) architectures have received considerable scientific attention due to their distinctive physicochemical attributes. These include larger surface area, enhanced charge transport, and reduced recombination of charge carriers. Within this category, highly ordered titania nanotube arrays are particularly notable. Their structure provides not only a greater surface area but also a direct pathway for charge carrier transport, which significantly enhances their functional performance compared to other morphologies such as nanorods [22].

4- Crystallographic Structure of Titanium Dioxide

Titanium dioxide (TiO_2) primarily crystallizes in three distinct polymorphs: rutile, anatase, and brookite. Among these, rutile represents the most thermodynamically stable phase under standard conditions, whereas anatase and brookite are metastable and can undergo transformation into rutile upon thermal treatment. From a crystallographic perspective, both rutile and anatase exhibit tetragonal symmetry, while brookite adopts an orthorhombic crystal structure. These distinct crystalline forms are illustrated in Figure 1-1 (a–c) [23].

In some general applications such as filtration, the crystalline phase may not be required. However, for specific functionalities—particularly photocatalytic or semiconducting purposes—the crystalline structure becomes essential. For example, the anatase phase is frequently utilized in photoactive pigments and photocatalytic systems due to its superior surface properties and reactivity, while the rutile phase is primarily applied in dielectric materials and high-temperature oxygen sensing devices. Owing to the distinct functional roles and structural characteristics of these two polymorphs, a more comprehensive analysis of anatase and rutile will be presented in the subsequent section [13].

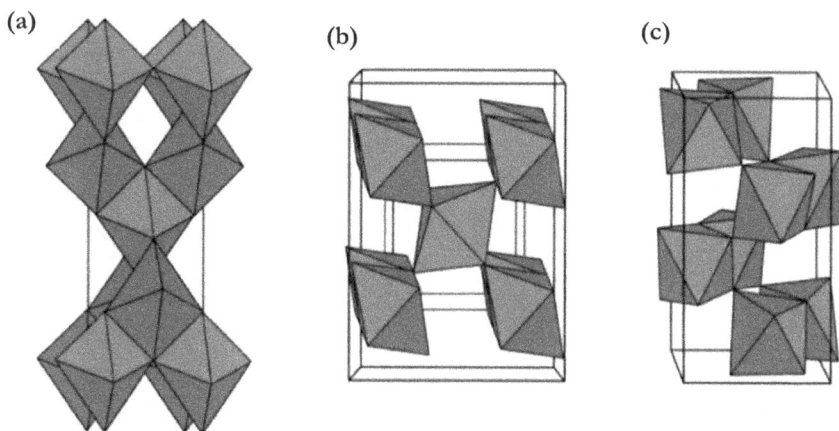

Figure 1-1: Crystal structures of (a) anatase, (b) rutile, and (c) brookite [24].

4-1 Rutile

Rutile crystallizes in a tetragonal lattice system and contains six atoms within each unit cell. In this arrangement, the TiO_6 octahedra are present in a slightly distorted form, deviating from perfect octahedral geometry. Rutile remains thermodynamically stable over a wide range of temperatures and pressures up to approximately 60 kilobars. Although rutile is generally considered photocatalytically inactive or weakly active, some studies suggest that its photocatalytic behavior is highly dependent on the synthesis conditions and preparation methods [25].

4-2 Anatase

Anatase also adopts a tetragonal crystal structure; however, its TiO_6 octahedra exhibit greater distortion than those found in the rutile phase. Owing to its superior electron mobility, lower dielectric constant, and reduced density, anatase is often favored over rutile and brookite for use in photovoltaic devices, particularly in solar cell technologies [26]. Titania is a wide bandgap semiconductor, with the bandgap energy values for anatase and rutile being approximately 3.20 eV and 3.02 eV, respectively [27]. From a structural perspective, both the anatase and rutile phases can be described as consisting of chains of edge-sharing TiO_6 octahedra, in which each titanium ion is coordinated by six oxide (O^{2-}) ions [28]. Variations in crystal lattice configuration, mass density, and electronic band structure among the polymorphic forms of TiO_2 give rise to notable differences in their physical and chemical behaviors [29].

CHAPTER 2: NANOSTRUCTURED TIO$_2$— ARCHITECTURES, ANODIC NANOTUBES, AND PROCESS–STRUCTURE RELATIONSHIPS

1- Introduction

Titanium dioxide (TiO$_2$) at the nanoscale exhibits properties that diverge significantly from those of its bulk counterpart, owing to quantum-confinement effects, high specific surface area, and abundant surface/defect states that govern charge transport and interfacial kinetics niumd [4,30]. Over the past two decades, rational control of TiO$_2$ into nanostructured architectures—such as nanoparticles, nanorods, nanowires, nanotubes, and mesoporous films—has enabled remarkable advances in photocatalysis, photoelectrochemistry, and energy storage by shortening carrier diffusion lengths and maximizing the density of active sites [4,30,31]. Among these geometries, one-dimensional (1D) arrays of anodic TiO$_2$ nanotubes (TNTs) are of particular importance. Their highly ordered vectorial morphology provides directional electron transport, mitigates grain-boundary trapping, and decouples light absorption from charge extraction, which are difficult to achieve in conventional particulate films [31,32].

Electrochemical anodization of Ti in fluoride-containing electrolytes represents a scalable, template-free technique for producing highly ordered TNT arrays with tunable inner and outer diameters, wall thicknesses, lengths, and intertube spacings. These geometrical parameters are dictated by a delicate interplay among field-assisted oxidation, field-assisted dissolution, and chemical etching. Consequently, substrate morphology, electrolyte composition, applied voltage, temperature, anodization duration, and even the choice of cathode material leave distinct imprints on nanotube architecture and, hence, on optoelectronic and photoelectrochemical performance [19,31,33].

This chapter introduces the general nanostructures of TiO$_2$ (Section 5), elaborates on the formation and unique properties of nanotubes (Section 6), and explores the influence of processing factors (Section 7 and subsections). By systematically analyzing how growth parameters—from mechanical and chemical surface treatments to electrolyte chemistry and thermal conditions—affect nanotube morphology, this chapter provides the scientific basis for the advanced modification strategies and applications discussed in subsequent chapters.

2- Nanostructures of Titanium Dioxide

Nanotechnology offers a set of advanced fabrication techniques that enable the design of materials with tailored structures to achieve desired functional properties. Materials synthesized at the nanoscale often exhibit novel behaviors that are not predictable based on their bulk counterparts [34]. In many applications, maximizing the specific surface area is crucial for enhancing performance. Moreover, in the nanoscale form, titanium dioxide demonstrates substrate-dependent behaviors not observed in its conventional oxide layer form.

Therefore, titania is frequently utilized in the form of nanoparticles or nanoscale architectures. Among various geometries, nanotubes and nanorods offer particularly favorable conditions for tuning both chemical and physical responses. As the dimensions shrink to the nanometer scale, not only does the surface area increase dramatically, but the electrical properties of the material also undergo substantial modifications [35].

3- Titanium Dioxide Nanotubes

The formation of titanium dioxide in nanotubular architectures at the nanoscale significantly increases the surface-to-volume ratio, thereby promoting more efficient electron transport and enhanced charge carrier separation. This structural configuration ultimately leads to a substantial reduction in the recombination rate of electron–hole pairs. In contrast, when the structure is not tubular—such as in thin films—electron movement becomes scattered, and the recombination rate of charge carriers increases due to carrier trapping at grain boundaries in the crystalline structure [36].

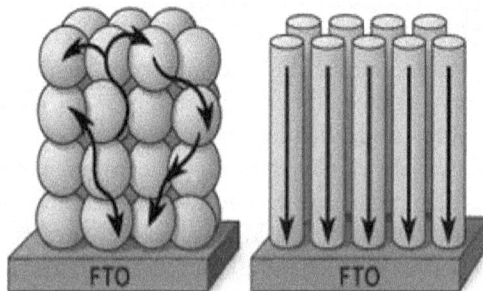

Figure 2-1: Electron transport in two different titanium dioxide structures: (a) TiO₂ nanoparticles, (b) TiO₂ nanotubes.

This one-dimensional nanostructure exhibits distinct electronic characteristics, most notably rapid electron transport, enhanced specific surface area, and improved electrical efficiency of the system. Although carbon nanotubes have been widely used in various applications, a broad class of transition metal oxides and sulfides has emerged, capable of engineered nanoscale growth into one-dimensional structures such as nanotubes, nanowires, and nanorods, each offering unique properties and functionalities. Among all transition metal oxides, titanium dioxide (TiO_2) has attracted the greatest attention, making it a highly researched compound in materials science. TiO_2 nanotubes can be synthesized across a wide range of pore diameters, wall thicknesses, lengths, and chemical compositions. These structural parameters influence their photocatalytic activity and light absorption capacity, leading to different levels of photon-to-current conversion efficiency. Moreover, TiO_2 nanotubes show promising potential in pharmaceutical and biomedical fields. For example, they can be used as agents to promote blood coagulation for hemorrhage control, as well as for catalyzing organic reactions and for detecting oxygen and hydrogen gases [37].

The fabrication of one-dimensional titania nanostructures can be achieved through a variety of methods, including the sol–gel technique, template-assisted synthesis [38], hydrothermal processes, and electrochemical methods.

4- The Influence of Various Factors on the Properties of Titanium Dioxide Nanotubes

The properties of titanium dioxide (TiO_2) nanotubes synthesized via the anodic oxidation method are significantly influenced by key geometrical parameters such as tube length, inner and outer diameters, wall thickness, and the spacing between adjacent nanotubes. These parameters are, in turn, governed by a range of processing conditions that must be precisely controlled to obtain optimal structural and functional characteristics. Critical factors requiring optimization before, during, and after the anodization process include the surface morphology of the substrate, the type and composition of the electrolyte, the applied voltage, anodization duration, temperature, electrolyte pH, electrolyte viscosity, and the material and configuration of the cathode used. Careful regulation of these parameters allows for the tailoring of nanotube architecture to meet specific performance requirements in various applications.

4-1 Effect of Titanium Substrate Morphology and Surface Treatments

Dabi and He investigated the influence of titanium substrate surface morphology on the growth behavior and final properties of titanium dioxide nanotubes. Their findings highlighted that the initial surface condition of the titanium substrate plays a critical role in determining the morphology of the resulting nanotube arrays. This section focuses on examining the impact of both mechanical and chemical surface treatments, which significantly affect the nucleation, growth uniformity, and structural order of the synthesized nanotubes [39].

4-2 Effect of Mechanical Surface Treatment on the Titanium Substrate

For the fabrication of titanium dioxide nanotubes, commercially available titanium foil with a typical thickness of approximately 250 μm is commonly utilized. Prior to the anodization process, the titanium substrate undergoes a cleaning procedure involving detergents, ethanol, toluene, and deionized water to remove surface contaminants and impurities. Although several commercial suppliers provide high-purity titanium foil, their surface morphologies may vary considerably. Investigations have shown that these foils typically exhibit surface cracks distributed across the substrate. The presence of such cracks results in the formation of vertical grooves and, consequently, regions devoid of material at certain depths below the surface layer [39].

In addition, submicron-scale surface inhomogeneities are often observed in the vicinity of crack sites, contributing to an increased substrate surface roughness. The presence of these cracks on the titanium substrate leads to greater morphological irregularities in the resulting nanotube arrays, frequently causing the formation of nanotube clusters and agglomerated domains.

The local morphology of the nanotubes in the vicinity of these crack regions has also been investigated. A high-magnification SEM image (Figure 2-2a) clearly demonstrates the substantial influence of substrate morphology on the early-stage growth of nanotubes. This effect is particularly pronounced at short anodization durations, where the nanotubes have not yet fully aligned. The clusters tend to form near the substrate crack lines, whereas a uniform nanotube distribution is typically observed on crack-free regions. Furthermore, it was observed that nanotubes located at the edges of cracks tend to intertwine, leading to the formation of cross-linked nanotubes, as shown in Figure 2-2b. These entangled structures have the potential to contribute to the development of nanotube clusters [39].

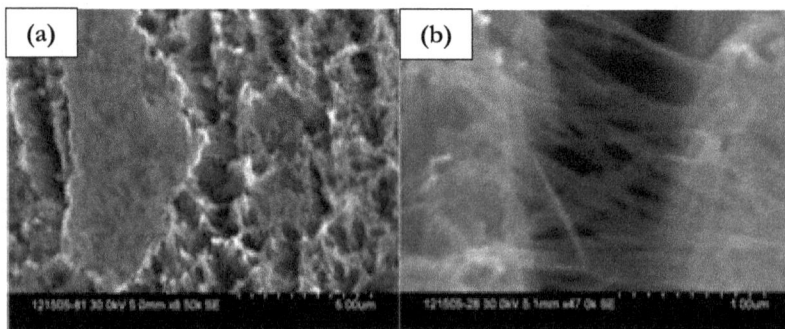

Figure 2-2: SEM images of TiO₂ nanotubes after 30 minutes of anodization: (a) Nanotube clusters formed along substrate crack lines; (b) Intertwined TiO₂ nanotubes observed at crack edges [39].

Findings from the anodization of commercially pure titanium underscore that a smooth surface on the substrate plays a critical role in the development of uniformly organized nanotubular morphologies. Both Han et al. [40] and Li et al. [41] have independently demonstrated that a two-step anodization technique can effectively yield highly uniform nanotube structures. In their approach, the initial anodization is conducted for a relatively short duration, followed by the complete removal of the primary nanotube layer. This process results in a significantly smoother titanium surface, which in turn facilitates the formation of well-aligned nanotube arrays during the second anodization stage. Kang et al. [42] employed an electropolishing method in which the titanium substrate undergoes electrochemical polishing prior to anodization, producing a smooth surface that enables the development of highly ordered nanotube structures. Although both techniques have proven effective in generating uniform nanotube arrays, they are considered less practical due to their time-consuming nature and higher processing costs. As an alternative to mitigate surface cracking, mechanical polishing using fine-grade sandpapers has been proposed. However, SEM analysis revealed that even the use of ultra-fine abrasive papers leaves behind microscale scratches on the surface. These surface defects lead to evident clustering and agglomeration of nanotubes. Therefore, it can be concluded that despite careful mechanical polishing, microscale roughness remains, which is unsuitable for the growth of highly ordered nanotube architectures [39].

4-3 The Effect of Chemical Treatment on the Titanium Substrate Surface

As previously discussed, mechanical polishing alone is insufficient to completely eliminate surface cracks. Therefore, chemical treatments can be employed to achieve a crack-free surface, one of which is chemical etching. In the experimental investigation conducted by Dabi and Han, titanium substrates were subjected to immersion in a 0.75 M hydrofluoric acid (HF) solution for varying durations ranging from 1 to 15 minutes. It was observed that complete removal of surface cracks was achieved after 10 minutes of etching. As depicted in Figure 2-3a, the scanning electron microscopy (SEM) image of the specimen treated for 5 minutes still exhibits distinct vertical cracks and surface fissures. In contrast, Figure 2-3b illustrates that these defects are entirely eliminated after 10 minutes; however, the substrate surface remains relatively rough. Extending the etching duration to 15 minutes results in an irregular and highly rough surface, as shown in Figure 2-3c [39]. A detailed inspection of the isolated pits formed following 10 minutes of etching, as illustrated in Figure 2-3d, indicates that these structures exhibit a smooth, concave morphology with average diameters ranging between 5 and 10 micrometers. These findings suggest that well-aligned and ordered TiO$_2$ nanotubes can be grown on such smooth, locally ordered surfaces. Moreover, the concave curvature of the etched pits may induce a slight bending of the nanotubes, which in turn helps prevent entanglement and promotes overall morphological uniformity in the resulting nanotube arrays [39].

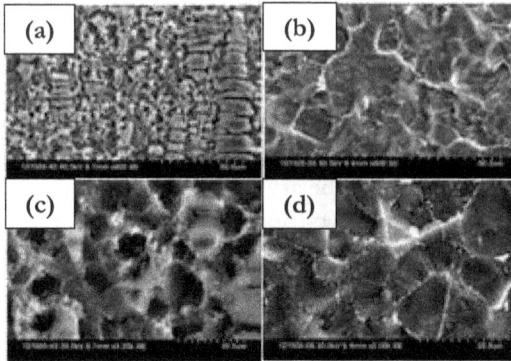

Figure 2-3: Scanning electron microscopy (SEM) images of titanium foil etched in a 0.75 M hydrofluoric acid (HF) solution for varying durations: (a) 5 minutes, (b) 10 minutes, (c) 15 minutes, and (d) 10 minutes under higher magnification conditions [39].

In order to substantiate this hypothesis, a 30-minute anodization process was carried out to produce short TiO_2 nanotubes on a titanium substrate that had been chemically etched in a 0.75 M HF solution for 10 minutes. The SEM image presented in Figure 2-4a illustrates the surface morphology of the nanotubes formed during this step, clea   demonstrating that the nanotubes closely replicate the geometry of the individual pits across the entire substrate. Furthermore, no signs of clustering or entanglement are observed at any location, confirming that this process can yield a uniformly ordered nanotube structure. However, the practical applicability of this method must also be demonstrated for the growth of longer nanotubes with long-range order. To investigate this, anodization was extended to 5 hours on the same pre-etched substrate, resulting in nanotubes approximately 20 μm in length, as shown in Figure 2-4b. Remarkably, even after prolonged anodization, the nanotubes maintain a well-ordered morphology with no evidence of clustering across the surface. These findings provide strong evidence that the morphology of TiO_2 nanotubes is highly influenced by the initial surface condition of the substrate. Moreover, they underscore the critical role of appropriate chemical pretreatment in promoting the formation of well-ordered nanotube arrays [39]. Numerous researchers have explored the influence of titanium substrate surface conditions on the morphology of anodically grown nanotubes. For instance, Peyghambardoust et al. [43] examined the effects of mechanical polishing and electropolishing on the degree of nanotube ordering. They demonstrated that despite similar surface roughness in both cases, mechanical polishing introduces specific surface textures on the titanium substrate that adversely affect the uniformity of nanotube growth during anodization.

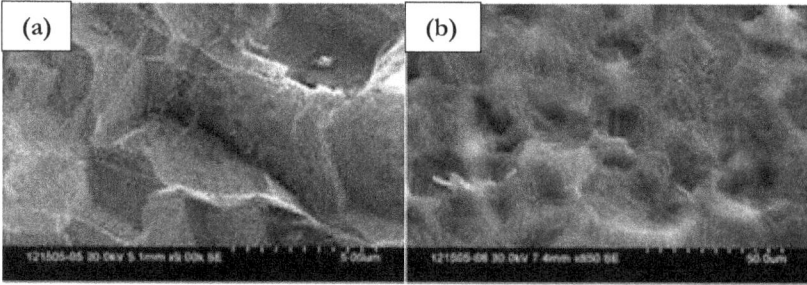

Figure 2-4: SEM images of TiO$_2$ nanotubes formed on titanium foil after anodization for (a) 30 minutes and (b) 5 hours [39].

4-4 Effect of Temperature and Voltage

Mor et al. [44]investigated the influence of anodization temperature on the wall thickness, length, and photovoltaic performance of TiO$_2$ nanotubes synthesized under different thermal conditions. In their study, anodization was conducted at a constant voltage of 10 V in an electrolyte composed of 0.5% HF and acetic acid, mixed in a 7:1 volume ratio, at four different temperatures: 5 °C, 25 °C, 35 °C, and 50 °C. Additional trials were conducted at 20 V under two temperature conditions: 5 °C and 25 °C.

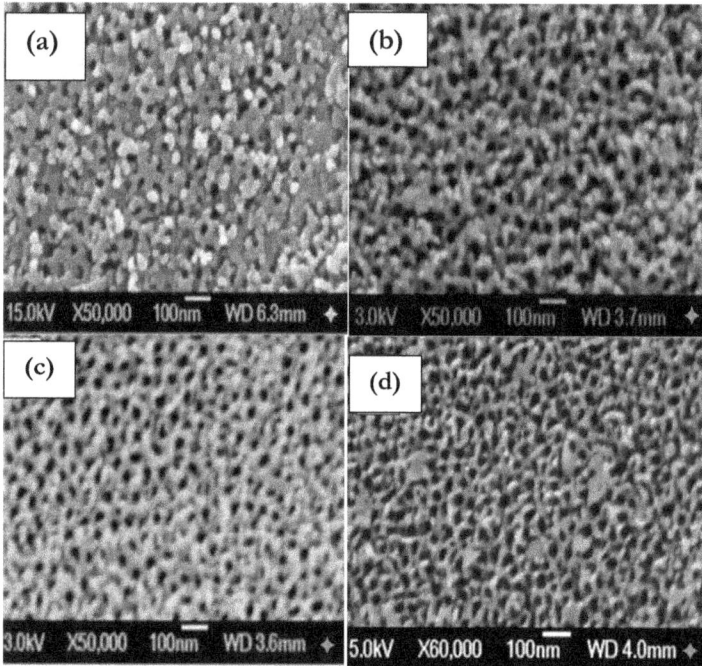

Figure 2-5: FESEM images of TiO$_2$ nanotubes anodized in 0.5% HF–acetic acid electrolyte at 10 V, synthesized at different temperatures: (a) 5 °C, (b) 25 °C, (c) 35 °C, and (d) 50 °C [44].

Figure 2-5 presents FESEM images illustrating the morphological characteristics of the nanotubes formed at (a) 5 °C, (b) 25 °C, (c) 35 °C, and (d) 50 °C. While the pore diameter remained constant at approximately 22 nm across all temperatures, notable variations were observed in both wall thickness and tube length. Specifically, the wall thickness increased significantly as the temperature decreased—from 9 nm at 50 °C to 34 nm at 5 °C. This thickening of the walls resulted in narrower inter-tubular gaps, progressively converting the distinct tubular architecture into a more interconnected nanoporous network. Similarly, tube length exhibited a positive correlation with decreasing temperature, increasing from 120 nm at 50 °C to 224 nm at 5 °C [44].

Figure 2-6 shows FESEM images of TiO₂ nanotubes anodized at 20 V at two temperatures: (a) 5 °C and (b) 25 °C. In both samples, the inner pore diameter measured 76 nm. However, the wall thickness increased from 17 nm at 25 °C to 27 nm at 5 °C, reinforcing the trend that lower anodization temperatures result in thicker nanotube walls [44].

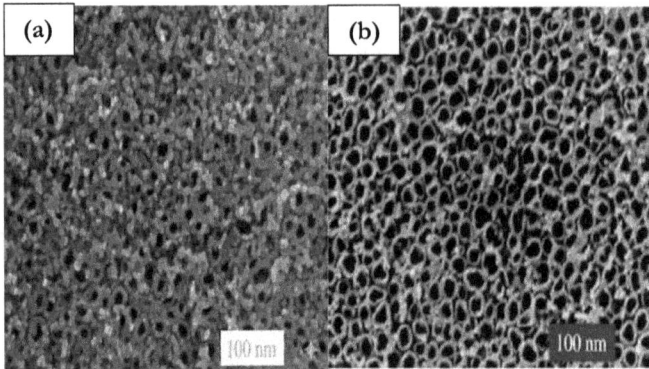

Figure 2-6: FESEM images of TiO₂ nanotubes anodized at 20 V under two temperature conditions: (a) 5 °C and (b) 25 °C [44].

Table 2-1 Changes in wall thickness and tube length of TiO₂ nanotubes with respect to anodization temperature at two applied voltages: 10 V and 20 V.

Table 2-1: Characteristics of TiO₂ nanotubes synthesized at different anodization temperatures [44].

anodization temp	Voltage 10V			Voltage 10V	
	wall thickness (nm)	length (nm)	inner diameter	wall thickness (nm)	inner diameter
5	34	224	22	27	76
25	24	176	22	17	76
35	13.5	156	22	-	76
50	9	120	22	-	76

The observed variations can be attributed to the underlying formation mechanism of TiO_2 nanotubes. In this process, the chemical dissolution of titanium oxide in an HF-containing electrolyte plays a central role, ultimately determining the maximum achievable tube length. Since wet chemical etching is strongly temperature-dependent—typically exhibiting an exponential increase in rate with temperature—the production of soluble ionic species becomes more pronounced at higher temperatures. Consequently, it can be inferred that fluoride-ion-assisted etching proceeds less aggressively at lower temperatures, thereby allowing for the growth of nanotubes with increased wall thickness and overall length. In contrast, the pore diameter remains largely unaffected by temperature fluctuations, as it is predominantly controlled by the applied anodization voltage rather than thermal conditions [44].

4-5 Effect of Cathode Material

The material of the cathode can significantly influence the electrochemical reactions occurring during anodization, as well as the rate at which they proceed. Alam and Grimes [45] explored the influence of various cathode materials on the structural and morphological characteristics of TiO_2 nanotubes formed through anodization. Their experiments were carried out at a fixed voltage of 20 V over a 10-hour period, using both aqueous and ethylene glycol-based electrolytes.

FESEM analysis of anodized titanium foil samples, prepared in an aqueous electrolyte containing 0.2M NH_4F and 0.1M H_3PO_4 at a constant voltage of 20V, demonstrated the impact of different platinum-group cathode materials—including platinum, palladium, and nickel—on nanotube morphology. The results indicated that nanotube arrays synthesized using nickel and palladium cathodes exhibited structural features closely resembling those obtained with platinum, with only minor variations in tube diameter and length [45]. Figure 2-7a displays the FESEM image of the sample anodized using a nickel cathode. For transition metals outside the platinum group, TiO_2 nanotubes formed using an iron cathode are shown in Figure 2-7b. When compared with those produced using cobalt and copper cathodes, it was observed that highly ordered nanotubes with varying lengths were obtained. Top-view images indicated that the tube diameter did not change proportionally with their length. Nanotube arrays synthesized using a tantalum cathode exhibited larger pore diameters compared to those formed with platinum. Conversely, the shortest nanotubes were obtained with a tungsten cathode, as illustrated in Figure 2-7c. For non-transition metal cathodes, nanotube arrays fabricated using carbon and tin electrodes demonstrated comparable dimensions. However, as shown in Figure 2-7d, the nanotubes synthesized with a carbon cathode exhibited a more uniform morphology. In contrast, the use of aluminum cathodes led to the formation of the shortest TiO_2 nanotubes, characterized by poor structural organization and low morphological order [45].

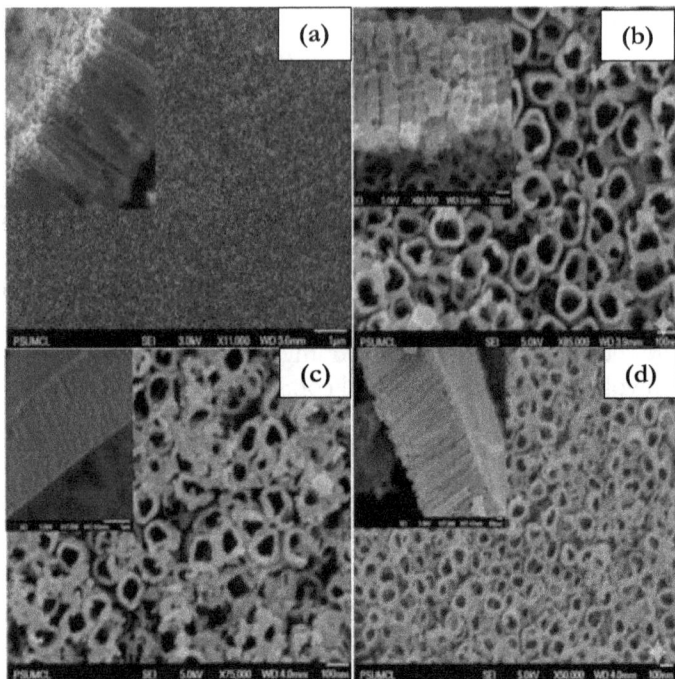

Figure 2-7: FESEM images of TiO$_2$ nanotubes anodized for 10 hours in an aqueous electrolyte at 20 V, using various cathode materials: (a) nickel, (b) iron, (c) tungsten, and (d) carbon [45].

Figure 2-8a presents FESEM images of titanium foil samples anodized in an ethylene glycol-based electrolyte containing 0.2 M NH$_4$F and 0.1 M H$_3$PO$_4$, under a constant voltage of 20 V, using platinum and nickel as cathode materials. These images reveal that the surfaces of the nanotube arrays are covered with collapsed fragments or debris. In contrast, as seen in Figure 2-8b, the arrays formed using a palladium cathode exhibit no such debris. A front-view SEM image of the nanotubes synthesized with the palladium cathode demonstrates a significantly higher degree of structural order compared to those fabricated with platinum or nickel cathodes [45].

Figure 2-8c displays FESEM images of TiO$_2$ nanotube arrays fabricated using an iron cathode, showing clean surfaces without any residual fragments or signs of structural collapse. Additionally, front-view images of nanotubes grown with cobalt, copper, tantalum, and tungsten cathodes demonstrate highly ordered arrays. However, top-view SEM images of these samples indicate that their surfaces are partially covered with adhered surface fragments, likely resulting from structural collapse or post-anodization residues [45].

Table 2-2: Influence of cathode material on the diameter and length of TiO_2 nanotubes [45].

Group	Cathode Material	Fabricated aqueous nanotubes in electrolytes		Fabricated EG nanotubes in electrolytes	
		Average diameter 5–7 (nm)	Average length 10 (nm)	Average diameter 5(nm)	Average length 10 (nm)
Pt-group elements	Ni	143	1200	65	1510
	Pd	134	1435	61	2500
	Pt	105	1520	65	1725
Non-Pt transition elements	Fe	99	2470	68	2000
	Co	135	1900	143	2100
	Cu	81	1265	83	1130
	Ta	140	1175	70	1300
	W	91	690	114	2400
Non-transition elements	C	143	1300	81	1220
	Al	96	570	85	1600
	Sn	147	1220	90	1060

Figure 2-8d illustrates TiO_2 nanotube arrays synthesized using a cobalt cathode. Nanotubes fabricated with carbon, tin, and aluminum cathodes demonstrated highly ordered architectures with surfaces largely free of residual particles. The morphology of the arrays produced with a carbon cathode is presented in Figure 2-8e [45].

Table 2-2 provides a summary of how different cathode materials affect the diameter and length of the synthesized TiO_2 nanotubes.

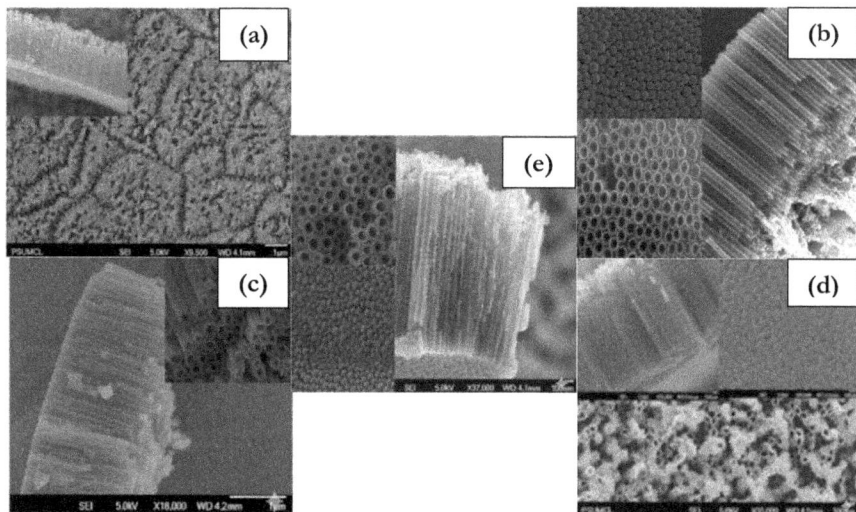

Figure 2-8: FESEM images of TiO_2 nanotubes synthesized via 10-hour anodization in an ethylene glycol-based electrolyte at 20 V, using various cathode materials: (a) platinum, (b) palladium, (c) iron, (d) cobalt, and (e) carbon [45].

7-6 Effect of Doped Elements

Asahi et al. examined the effects of different dopant elements—including phosphorus (P), fluorine (F), nitrogen (N), carbon (C), and sulfur (S)—on the photocatalytic performance of TiO_2 thin films. Their findings demonstrated that among the elements studied, nitrogen had the most significant effect on enhancing TiO_2 photocatalytic performance. This is attributed to the fact that nitrogen's p-orbitals, when combined with the oxygen 2p states, lead to a narrowing of the bandgap. Nitrogen atoms can be introduced into the TiO_2 crystal lattice through either substitutional or interstitial doping mechanisms. The substitutional nitrogen, which replaces some of the oxygen atoms, is considered the more effective form. This form, denoted as β-N and identified in XPS analysis with a binding energy of approximately 396 eV (Figure 2-9a), is primarily responsible for the observed photocatalytic improvement. Figure 2-9b shows the variation in light absorption after 10 hours of irradiation as a function of the proportion of substitutional nitrogen relative to the total nitrogen content—kept at approximately 1 atomic percent across all samples. As depicted in the figure 2-9, the optimal concentration of substitutional nitrogen is around 0.25 atomic percent. Although sulfur is also a potentially effective dopant, its relatively large ionic radius makes its incorporation into the TiO_2 crystal lattice more difficult [46].

Maleki and colleagues investigated the photocatalytic properties of TiO_2 nanotubes co-doped with nickel, carbon, and nitrogen. In this investigation, TiO_2 nanotubes were fabricated by anodizing high-purity titanium foil in an ethylene glycol-based electrolyte containing 0.42 wt% NH_4F and 1.85 wt% H_2O. The anodization process was carried out at a constant voltage of 40 V for 2 hours under ambient conditions.

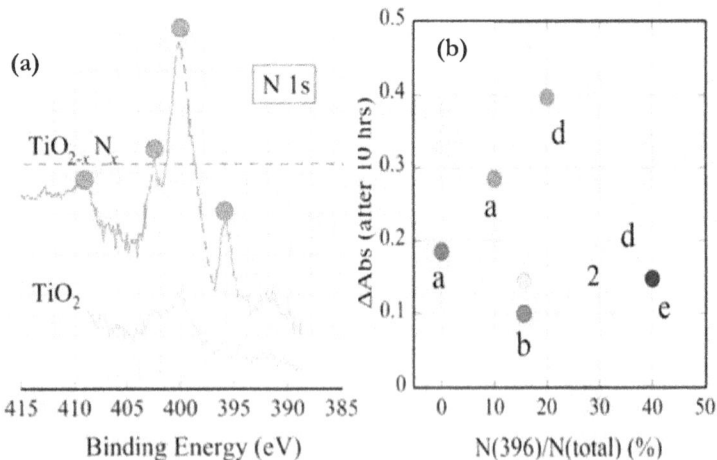

Figure 2-9: (a) XPS analysis of N1s for TiO_2 before and after doping; (b) Relationship between light absorption after 10 hours of irradiation and the proportion of substitutional nitrogen relative to total doped nitrogen [46].

For one-step co-doping with Ni, C, and N, varying concentrations (0.008, 0.015, 0.028, 0.045, and 0.07 wt%) of potassium tetracyanonickelate hydrate $(K_2Ni(CN)_4 \cdot xH_2O)$ were introduced directly into the electrolyte prior to anodization. Post-synthesis, the nanotube arrays underwent thermal annealing at 500 °C for 2.5 hours to promote crystallization of the initially amorphous TiO_2 phase [47].

The FESEM results, shown in Figure 2-10, indicate that doping elements had no significant effect on the inner diameter or wall thickness of the TiO_2 nanotubes. In all three samples, the nanotubes exhibited a consistent inner diameter of 75 ± 7 nm. However, an increase in dopant concentration resulted in a reduction in nanotube length—from 10.2 nm for undoped TiO_2 to 7.4 nm for the sample doped with 0.045 wt% of $K_2Ni(CN)_4 \cdot xH_2O$. Furthermore, EDX analysis, as shown in Figure 2-11, verifies the successful incorporation of nickel, nitrogen, and carbon within the doped TiO_2 samples [47].

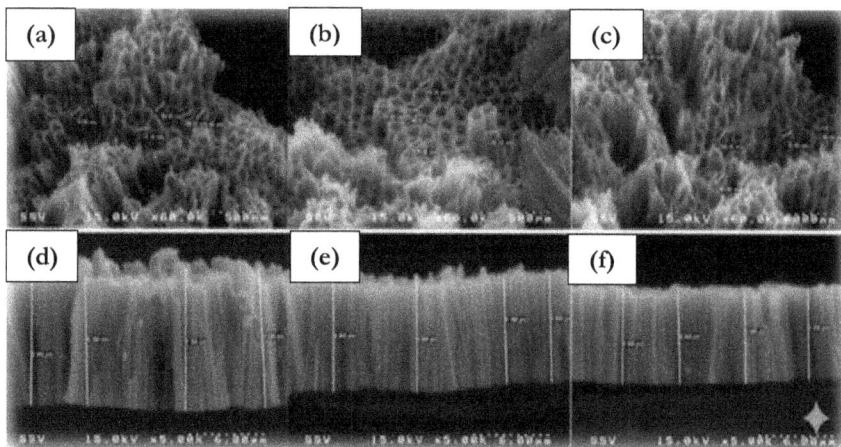

Figure 2-10: Top- and side-view FESEM images of TiO_2 nanotubes: (a, d) undoped, (b, e) doped with 0.015 wt% $K_2Ni(CN)_4$, (c, f) doped with 0.045 wt% $K_2Ni(CN)_4$ [47].

XRD analysis indicates that elevating the concentration of $K_2Ni(CN)_4$ in the electrolyte from 0.008 wt% to 0.045 wt% leads to improved crystallinity of the TiO_2 nanotubes, likely as a result of effective incorporation of dopant atoms into the crystal lattice. However, further increasing the salt concentration to 0.07 wt% leads to a noticeable decline in crystallinity. This reduction in the anatase phase intensity can be attributed to two factors: a significant decrease in nanotube length at higher dopant levels, and increased X-ray penetration into the underlying titanium substrate, as evidenced by the rising intensity ratio of titanium to TiO_2 peaks. Figure 2-12 presents the XRD patterns for all samples. The inset graph in the same figure shows that, with increasing $K_2Ni(CN)_4$ concentration, the main diffraction peak of the anatase phase shifts from 37.8°

to 38.3°, indicating slight lattice distortion due to dopant incorporation [47].

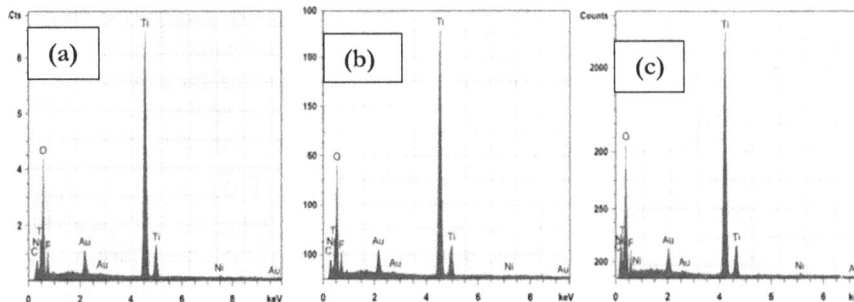

Figure 2-11: EDX elemental analysis results of TiO_2 nanotube samples: (a) undoped, (b) doped with 0.015 wt% $K_2Ni(CN)_4$, and (c) doped with 0.045 wt% $K_2Ni(CN)_4$ [47].

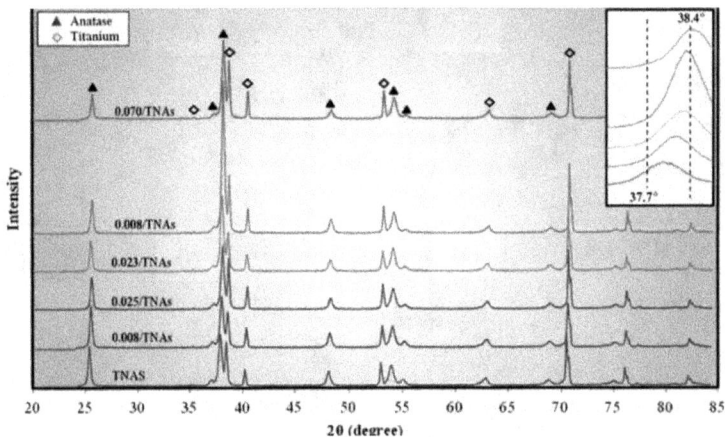

Figure 2-12: XRD patterns of TiO_2 nanotube samples prepared in electrolytes with varying concentrations of $K_2Ni(CN)_4$; inset (top right) shows the effect of salt concentration on the interplanar spacing (diffraction angle) of the anatase phase [47].

XPS analysis results for the N1s, O1s, Ni2p, Ti2p, and C1s peaks of the TiO_2 nanotube sample prepared in an electrolyte containing 0.045 wt% $K_2Ni(CN)_4$ are presented in Figure 2-13. As shown in Figure 2-13a, the dominant chemical state of nickel in the sample is Ni^{2+}, indicating that Ni atoms have substituted Ti atoms within the TiO_2 lattice. The nickel concentration determined by XPS was 3.8%, whereas the corresponding value from EDX analysis was approximately 1%. This variation implies that Ni ions are primarily integrated into the outermost nanometers of the TiO_2 layer during nanotube formation. Such non-uniform distribution is likely due to the preferential interaction of larger anions with the porous anodic oxide matrix [47].

The XPS result for titanium (Figure 2-13b) confirms that Ti^{4+} exists predominantly in the oxidation state, corresponding to TiO_2. Figure 2-13c, presenting the XPS spectrum of the O1s region, reveals three distinct peaks at binding energies of 529.87 eV, 531.27 eV, and 532.5 eV. These peaks correspond to lattice oxygen associated with Ti–O bonds, oxygen vacancy states, and surface hydroxyl groups, respectively. The presence of oxygen vacancies supports the substitution of Ti^{4+} by Ni^{2+} ions in the TiO_2 lattice, as this substitution creates local charge imbalance that is compensated by oxygen deficiency [47].

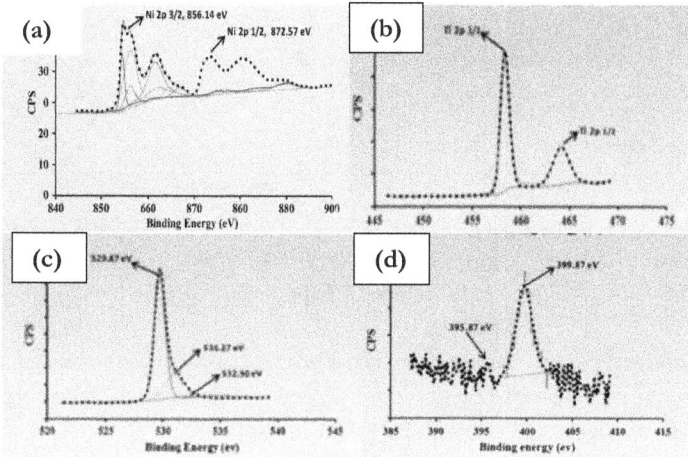

Figure 2-13: XPS spectra of TiO_2 nanotubes synthesized in an ethylene glycol-based electrolyte with 0.045 wt% $K_2Ni(CN)_4$, showing: (a) Ni 2p, (b) Ti 2p, (c) O 1s, (d) N 1s, and (e) C 1s regions [47].

In Figure 2-13d, the XPS spectrum of the N 1s region exhibits two distinct peaks. The peak at 395.87 eV is attributed to N^{3-} species (β-N), typically linked to titanium oxynitride ($TiO_{2-x}N_x$) phases embedded within the anatase lattice. A second, higher-energy peak at 399.97 eV corresponds to substitutional nitrogen atoms replacing lattice oxygen [47].

Figure 2-14 displays the linear sweep voltammetry (LSV) curves of different TiO_2 photoanodes under visible light illumination. As shown, all Ni–C–N co-doped TiO_2 nanotube electrodes exhibit significantly higher photocurrent densities compared to undoped TiO_2 over the entire range of applied bias voltages. This enhancement in photocurrent can be attributed to two primary mechanisms. Firstly, the incorporation of nickel ions—capable of existing in multiple oxidation states—effectively reduces the recombination of photogenerated electron–hole pairs, thereby improving photocatalytic efficiency. Secondly, the substitution of oxygen atoms with nitrogen or carbon may lead to the formation of hybridized energy states comprising C 2p, O 2p, and N 2p orbitals. Given that the 2p orbitals of nitrogen and carbon lie at higher energy levels than the TiO_2 valence band, they can be activated under visible

21

light, resulting in enhanced charge carrier generation and superior photoelectrochemical performance.

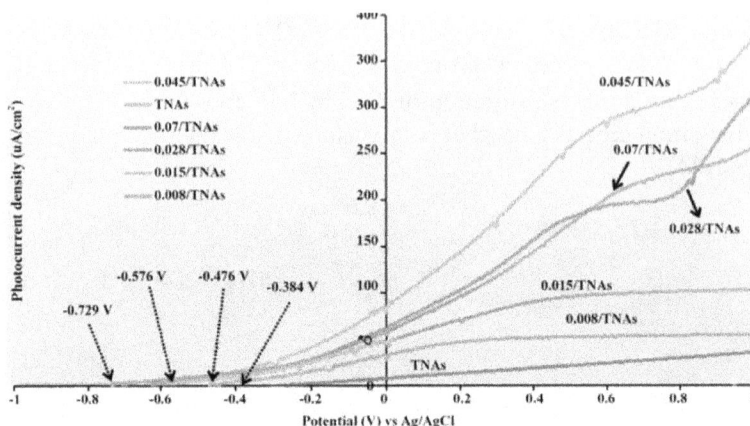

Figure 2-14: Photocurrent density versus applied bias voltage for undoped and Ni–C–N co-doped TiO$_2$ nanotube arrays under visible light, measured at a scan rate of 20 mV/s [47].

The reduced photocurrent observed in the sample prepared with 0.07 wt% K$_2$Ni(CN)$_4$ in the electrolyte is likely due to excessive ionic concentration, which may induce structural defects. These defects serve as recombination centers for photogenerated electron–hole pairs, ultimately diminishing the photocatalytic efficiency of the system [47].

CHAPTER 3: MODIFICATION STRATEGIES FOR TIO₂ NANOTUBES—DEFECT ENGINEERING, DOPING, AND ADVANCED FUNCTIONALIZATION

1- Introduction

Although pristine titanium dioxide nanotubes (TNTs) possess a highly ordered morphology and large surface area, their intrinsic electronic conductivity and specific capacitance remain relatively low, which severely limits their electrochemical and photoelectrochemical efficiency [30,31]. To overcome these limitations, extensive research has been devoted to modification strategies that tailor the structural, electronic, and surface properties of TNTs. Among the most widely adopted approaches are thermal treatments, defect engineering through oxygen vacancies, incorporation of noble metals or conductive oxides, and particularly, elemental doping with metal and non-metal species [46,48].

Non-metal doping has emerged as a particularly effective pathway, as substitutional or interstitial incorporation of light elements such as nitrogen, carbon, sulfur, and fluorine can narrow the TiO_2 band gap and extend its light absorption from the ultraviolet into the visible range [4]. This band-structure engineering enhances charge-carrier separation and suppresses recombination, thereby significantly improving photocatalytic and photoelectrochemical responses. Similarly, self-doped or "black TiO_2" nanotubes, generated by electrochemical or thermal reduction, exploit oxygen vacancies and Ti^{3+} states to achieve quasi-metallic conductivity without the drawbacks commonly associated with extrinsic dopants [49].

In addition to doping, hybridization of TiO_2 with graphene, carbon nanotubes, or conductive polymers provides synergetic benefits by creating efficient electron-transport channels and large interfacial areas, thereby enhancing both photoelectrochemical conversion efficiency and electrochemical capacitance [50]. Collectively, these modification strategies illustrate the versatility of TNTs as a platform material that can be tuned for a broad range of advanced applications—from photocatalysis and solar energy conversion to supercapacitors, sensors, and biomedical systems.

This chapter surveys the major modification approaches for TiO_2 nanotubes, with a particular emphasis on non-metal doping and self-doping strategies, while also highlighting the critical role of defect states and hybrid architectures in optimizing performance. By establishing clear structure–property relationships, these insights pave the way for designing next-generation TiO_2-based nanostructures with superior functionality across energy, environmental, and biomedical domains.

2- Modification of TiO_2 Nanotubes

Although titanium dioxide nanotubes (TNTs) possess an excellent morphological structure and high specific surface area, their electrochemical capacitance remains relatively low. In most studies, the specific capacitance of TNTs has been reported to be below $1\,mF\,cm^{-2}$, which is comparable to conventional electric double-layer capacitors. This limitation is primarily attributed to the inherently poor electrical conductivity of pristine TNTs [51].

To overcome this issue, researchers have focused on enhancing the electrical conductivity of TiO_2 electrodes through various strategies. These include the incorporation of impurities such as metal oxides, noble metals, non-metallic conductive materials, and self-doped TiO_2 structures [52]. Introducing oxygen vacancies (e.g., in Au/TiO_2 composites) [51] and employing conductive polymers [53] have also been shown to improve charge transport characteristics.

The enhancement of capacitance is generally achieved by increasing the surface area for ion adsorption, the density of charge carriers, and overall electrical conductivity [54]. Several modification strategies have been explored to enhance the capacitive behavior of TiO_2 nanotubes (TNTs), such as thermal annealing, nitrogen incorporation, elemental doping, plasma treatment, and electrochemical approaches [55]. Among these methods, elemental doping has garnered significant interest owing to its adaptability and proven effectiveness. The dopants can be either metallic (e.g., zirconium, iron, gold, chromium) or non-metallic (e.g., silicon, fluorine, carbon, nitrogen). Doping with both metallic [56,57] and especially non-metallic elements [58,59] has emerged as a widely adopted strategy for tailoring the structural and electronic properties of TNTs to enhance their electrochemical behavior.

2-1 Modification of TiO_2 Nanotubes via Non-Metal Doping

The crystal structure of a material is strongly governed by the relative sizes of cations and anions within the lattice. Given the disparities in ionic radii and charge states, substituting Ti^{4+} in the TiO_2 framework is typically more feasible than replacing O^{2-} with an alternative anionic species. However, studies have shown that doping TiO_2 with transition metal cations, while potentially improving thermal stability, tends to increase the optical bandgap and reduce charge carrier lifetime. These effects ultimately lead to a decrease in the material's photoelectrochemical efficiency, thereby limiting its light-to-energy conversion performance.

To address these drawbacks, anionic doping—particularly with non-metal elements—has gained widespread attention as a more effective strategy for enhancing the capacitive and photocatalytic properties of TiO_2 [60]. This strategy entails the substitution of lattice oxygen atoms with non-metallic dopants such as nitrogen [61], carbon [62], sulfur [63], and others. Incorporation of these elements alters the valence band structure of TiO_2 by introducing acceptor states above the valence band maximum, thereby enhancing visible-light absorption and promoting more efficient charge carrier separation. This band structure engineering is especially effective when applied to TiO_2 nanotubes (TNTs), allowing for the creation of high-performance electrode materials for photoelectrochemical applications.

2-2 Sulfur Doping of TiO₂

Sulfur-doped TiO_{2-x} with a core–shell nanostructure can be synthesized through aluminum reduction followed by thermal treatment in a hydrogen sulfide (H_2S) atmosphere. During the reduction step in a vacuum furnace, a disordered outer shell is formed, characterized by a high concentration of oxygen vacancies.

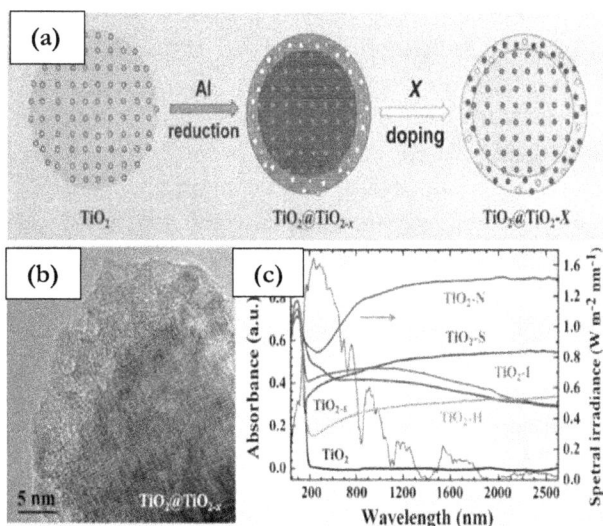

Figure 3-1: (a) Schematic of the synthesis and non-metal doping of core–shell TiO_{2-x}; (b) HRTEM image of the core–shell structure; (c) UV-Vis DRS spectra of TiO_{2-x} doped with H, N, S, and I [65].

These vacancies are subsequently occupied by S^{2-} anions, resulting in band gap narrowing and enhanced absorption in the visible light region. Similar to nitrogen doping, the narrowing of the band gap in sulfur-doped TiO_{2-x} is attributed to the hybridization of S 3p orbitals with impurity states linked to Ti^{3+} centers. Consequently, the absorption edge shifts from the ultraviolet (UV) region toward the visible and even near-infrared (NIR) spectrum [64].

2-3 Fluorine Doping of TiO₂

Fluorine doping of TiO_{2-x} samples is typically carried out through a vacuum thermal treatment process using a mixture of TiO_2/MCF and NH_4F. Similar to nitrogen doping, fluorine atoms occupy the oxygen vacancies within the TiO_{2-x} lattice during the heat treatment under vacuum conditions. Due to their strong electrophilic nature, the incorporation of F atoms leads to the formation of Ti^{3+}–F paramagnetic centers [66].

2-4 Carbon Doping of TiO₂

Due to the diverse properties of carbon-based materials, integrating carbon with TiO_{2-x} contributes to the development of new functionalities and enhanced performance. Graphene, a prototypical two-dimensional material originating from graphite, is composed of a single layer of carbon atoms arranged in a hexagonal lattice. It serves as the basic structural unit for all other graphitic forms. Graphene-based composites, owing to their exceptional electrical conductivity and unique structure, can serve as specialized support frameworks that promote charge separation and improve the migration of photoexcited electrons away from TiO_2 [67].

Graphene oxide (GO) sheets functionalized with TiO_2 nanocrystals (NCs) containing Ti^{3+} centers and oxygen vacancies (O_v) can be synthesized via a hydrothermal process followed by vacuum annealing. During this process, the TiO_2 NCs are functionalized with 3-aminopropyltriethoxysilane (APTES), introducing positively charged amine groups on their surface. Negatively charged GO sheets then electrostatically interact with the TiO_2 surface and bind effectively. High-resolution transmission electron microscopy (HRTEM) confirms the presence of a distinctive core–shell architecture within the TiO_2 nanocrystal/reduced graphene oxide (RGO) composite, which facilitates the generation of Ti^{3+} species and oxygen vacancy (O_v) defects. These structural alterations create impurity states below the conduction band, effectively shifting the light absorption edge toward the visible region. Consequently, photogenerated electrons can be efficiently transferred from the reduced TiO_2 to graphene, preventing their recombination with holes and enhancing photocatalytic performance [67].

2-5 Self-Doped TiO₂ Nanotubes

In semiconductor materials, impurities frequently serve as recombination centers for charge carriers. Although doping can effectively improve the electronic properties of TiO_2, the complexity and sensitivity of the doping process present significant practical challenges. Recent studies have shown that the introduction of intrinsic crystal defects in TiO_2 can generate vacancy-induced energy states located below the conduction band. These defect states enable visible-light absorption while avoiding the recombination issues often associated with extrinsic dopants [68]. In essence, electrochemical self-doping has emerged as a promising strategy for enhancing the electronic properties of TiO_2 [69].

One widely used technique to achieve self-doping is cathodization, where TiO_2 nanotubes are used as the cathode in an electrolyte solution. During this process, the color of the nanotubes changes from gray to blue-black. This transformation is attributed to the electrochemical reduction of multiple Ti(IV) centers to Ti(III) within the TiO_2 nanotube framework. The associated charge compensation is facilitated by the uptake of protons (H^+) from the solution. The observed color change is associated with the formation of sequential impurity states linked to Ti^{3+} centers, located approximately 0.3 to 0.8 eV below the conduction band Figure 3-2 schematically illustrates the energy band diagram of reduced and self-doped TiO_2. The increased electrical conductivity of self-doped TiO_2 nanotubes does not result from bandgap narrowing. Instead, it is governed by the position of the Fermi level (EF), which controls the conductivity of the semiconductor.

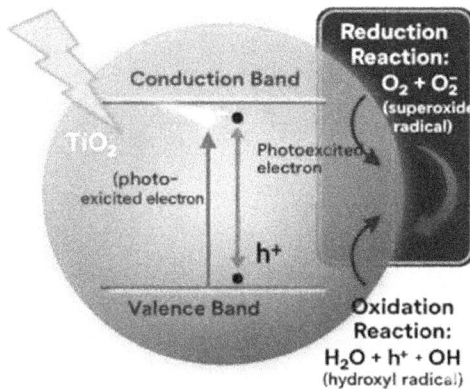

Figure 3-2: Schematic representation of the energy band structure of reduced TiO_2 [70].

At sufficiently high donor state densities, the Fermi level (EF) shifts above the conduction band edge (EC), transforming the material into a heavily doped n-type semiconductor with quasi-metallic behavior. Theoretical calculations have shown that, in black TiO_2 nanotubes, the Fermi level lies above the conduction band minimum, classifying them as reduced and self-doped TiO$_2$. Under these conditions, the energy states between the Fermi level (EF) and the conduction band edge (EC) become substantially populated with electrons, leading to increased electron density in the conduction band and enhanced electrical conductivity [70].

2-6 N-Doped TiO$_2$

Nitrogen is the most widely used non-metal dopant in TiO$_2$, primarily due to its small atomic radius and ionization energy, which are similar to those of oxygen [71,72].

Table 3-1: Methods for introducing nitrogen into TiO$_2$ and relevant nitrogen sources.

methods of nitrogen doping and source of nitrogen	References
Addition of N source to the TiO$_2$ precursor solution - Tetramethyl-ethylene-diamine	[73]
Electrochemical- Diethylenetriamine, ethylenediamine, hydrazine	[73–76]
Electrochemical anodization- Ammonium fluoride	[77]
plasma nitriding in N$_2$ atmosphere	[78,79]
ion implantation, heat treatment at 450 °C	[80,81]
decomposition of pure ammonia (NH$_3$) gas at 550 °C	[82]
heat treatment in N$_2$ atmosphere	[83]
electrochemical doping (NH$_4$F and CH$_3$NO) + heat treatment at 400 °C	[84]
electrochemical doping (NH$_4$Cl) + heat treatment at 450 °C	[85]
electrochemical doping (trimethylamine TEA) + heat treatment at 450 °C	[86]
electrochemical doping (urea) + heat treatment at 400 °C	[87]
electrochemical doping (various concentrations of urea) + heat treatment at 450 °C	[88]
electrochemical doping (various kinds of amines: DETA, TEA, EDA, urea)	[16]
hydrothermal method at 120 °C (trimethylamine)	[89]
wet immersion in NH$_3$ solution + heat treatment(450-700 °C)	[90,91]

Mahy et al. [92] conducted a comparative investigation of N-doped TiO$_2$ photocatalysts synthesized via an aqueous sol–gel route, employing titanium(IV) isopropoxide (TTIP) as the titanium precursor and three distinct nitrogen sources: urea, ethylenediamine (EDA), and trimethylamine (TEA), across various N/Ti molar ratios. The resulting materials exhibited enhanced visible-light activity. Among them, photocatalysts doped with urea and TEA yielded anatase–brookite phase TiO_2 nanoparticles with specific surface areas ranging between 200 and 275 $m^2 \cdot g^{-1}$. In contrast, EDA-modified samples developed a rutile phase, with surface areas between 185 and 240 $m^2 \cdot g^{-1}$. X-ray photoelectron spectroscopy (XPS) confirmed nitrogen incorporation through Ti–O–N bonding, which accounted for enhanced visible-light absorption.

According to literature data [93], nitrogen can be incorporated into the TiO_2 nanotube (NT) lattice via physical, chemical, or electrochemical methods, resulting in characteristic XPS signals. Nitrogen-related XPS peaks within the 399.0–403.0 eV range are typically associated with chemisorbed surface species or interstitial/substitutional nitrogen, depending on the doping method employed. In contrast, peaks appearing between 396.0 and 398.0 eV are indicative of substitutional nitrogen, corresponding to Ti–N bonds analogous to those found in titanium nitride (TiN) phases.

Under visible-light irradiation, the photocatalytic activity of TiO_2 is predominantly influenced by substitutional (N_s) and interstitial (N_i) nitrogen dopants. These species introduce mid-gap energy states that enable electron excitation from the impurity levels to the conduction band (CB), thereby enhancing photocatalytic performance. These electrons may subsequently be trapped by oxygen vacancies (O_v), improving charge separation in an energetically favorable process. However, undesirable recombination may occur between electrons in the CB or O_v states and holes at nitrogen impurity levels. Alternatively, photoexcited electrons may reduce molecular oxygen (O_2) to superoxide radicals ($O_2^{\bullet -}$), while photogenerated holes can oxidize water molecules or hydroxide ions to produce hydroxyl radicals (•OH). Both reactive species contribute significantly to the degradation of pollutants under visible-light irradiation [94,95].

CHAPTER 4: NITROGEN-DOPED TIO$_2$ NANOTUBE ARRAYS—PROPERTIES, CHARGE-TRANSPORT METRICS, AND SYNTHESIS PATHWAYS

1- Introduction

Nitrogen-doped TiO$_2$ nanotube arrays (N-TNTs) combine the architectural advantages of one-dimensional anodic tubes—vectorial electron transport, high interfacial area, and reduced grain-boundary trapping—with band-structure tailoring via anion doping. Compared with pristine TNTs, N-TNTs typically exhibit enhanced visible-light responsiveness and improved charge-separation efficiency due to N-derived states that narrow the apparent band gap and facilitate carrier generation under sub-UV irradiation [4,30,31,46]. The resulting performance gains manifest across photocatalysis and photoelectrochemistry, where faster interfacial kinetics and lower recombination rates are evidenced by reduced charge-transfer resistances (R$_{ct}$) and larger photocurrent densities [89,96].

Thermal processing strongly modulates structure and properties. Low-to-moderate annealing ($\approx$450 °C) crystallizes amorphous tubes to anatase while preserving the ordered morphology, whereas higher temperatures ($\geq$600–700 °C) drive wall thickening, rutile formation, and eventual collapse of the nanotubular architecture—outcomes that must be balanced against gains in crystallinity and mobility [97–99]. Electrochemical impedance spectroscopy (EIS) and Nyquist analyses provide quantitative handles on interfacial processes: N-TNTs commonly show smaller semicircles and lower R$_{ct}$ than undoped TNTs under illumination, consistent with improved carrier separation and accelerated interfacial charge transfer [89,96].

Multiple synthesis pathways enable control over N content, distribution, and local bonding environments—each with distinct trade-offs in morphology retention and defect chemistry. Routes include (i) anodization + nitrogen introduction (e.g., post-annealing in N$_2$/NH$_3$ or in-electrolyte N sources)

[85,87,100] (ii) direct thermal annealing in N-containing atmospheres [100], (iii) ion implantation to place N near O sites with minimal thermal load [87], and (iv) plasma-assisted nitriding that activates reactive N species at relatively low substrate temperatures [101,102]. Across these methods, careful coupling of substrate state, electrolyte chemistry, voltage/temperature profiles, and post-treatments is essential to preserve long-range tube order while establishing substitutional/interstitial N that optimizes optoelectronic response. This chapter synthesizes the structure–property relationships of N-TNTs and maps processing windows to targeted performance metrics, providing a practical template for designing next-generation TiO_2 nanoarchitectures for energy, environmental, and sensing applications.

2- Properties of N-TiO₂ Nanotube Arrays

Sun et al. who investigated nitrogen-doped TiO_2 nanotubes (N-TiO_2), reported that the morphology of doped and undoped nanotube arrays annealed at 450 °C remained highly similar. This observation suggests that low-temperature annealing has a negligible effect on nanotube morphology. As depicted in Figures 4-1a and b, the TiO_2 nanotubes exhibited an average diameter of approximately 80 nm and a wall thickness of around 15 nm [89].

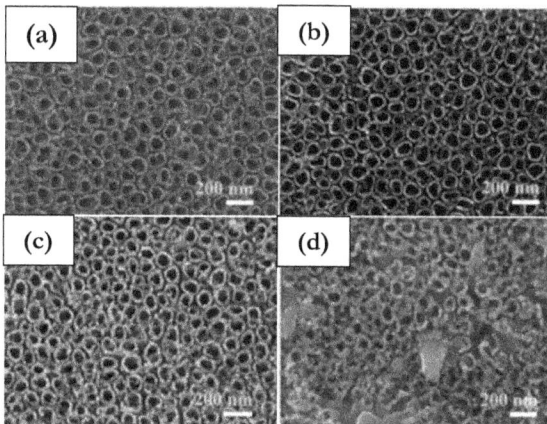

Figure 4-1:(a-c) SEM image of TiO_2 nanotube arrays annealed at 450 °C; (b–d) SEM images of N-doped TiO_2 nanotubes annealed at 450 °C, 600 °C, and 700 °C, respectively [91].

However, when the annealing temperature was raised to 600 °C, the N-doped TiO_2 samples exhibited a slight reduction in tube diameter accompanied by an increase in wall thickness, as shown in Figure 4-1c. Additionally, enhanced anatase crystal growth and a phase transition from anatase to rutile in undoped TiO_2 nanotubes were noted. Upon further increasing the temperature to 700 °C, the well-ordered nanotube architecture deteriorated, indicating a structural collapse of the array, as illustrated in Figure 4-1d. These findings highlight the thermal sensitivity of TiO_2 nanotube arrays and the importance of optimized annealing conditions to preserve their nanostructure while enhancing crystallinity [91].

The phase transition associated with variations in surface morphology and structural features of TiO_2 nanotubes was examined through X-ray diffraction (XRD) analysis. Figure 4-2 presents the XRD patterns of both undoped and nitrogen-doped titania nanotube films [91].

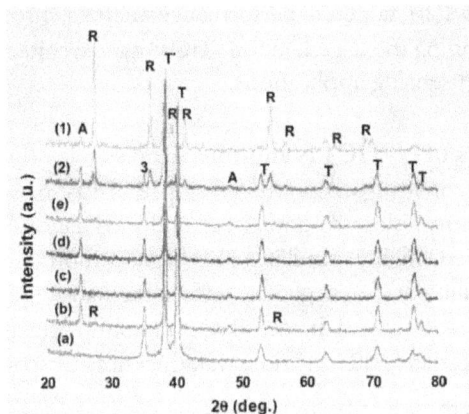

Figure 4-2: XRD patterns of (a) as-prepared titania nanotubes before annealing; (b) nanotubes annealed at 450 °C; and nitrogen-doped TiO_2 (N-TiO_2) nanotubes annealed at (c) 300 °C, (d) 450 °C, (e) 500 °C, and (f) 700 °C [91].

To investigate the electrochemical properties of nitrogen-doped TiO_2 nanotubes, a powerful and effective tool known as electrochemical impedance spectroscopy (EIS) is employed. This method is commonly employed to investigate electron transfer dynamics at the solid–liquid interface. Figure 4-3 displays Nyquist plots derived from electrochemical impedance spectroscopy (EIS) measurements performed in a 0.1 M aqueous Na_2SO_4 solution under visible-light illumination for both N-doped and undoped TiO_2 nanotube samples. The reduced semicircle diameter observed in the Nyquist plot for the nitrogen-doped TiO_2 sample, relative to the undoped counterpart, suggests that nitrogen incorporation effectively promotes the separation of photogenerated electron–hole pairs. The corresponding equivalent circuit model is depicted in Figure 4-3, and the fitted parameters are summarized in Table 4-1. The interfacial charge transfer resistance (R_{ct}) of the nitrogen-doped sample is noticeably lower than that of the undoped nanotubes. This lower resistance suggests a superior ability of the nitrogen-doped TiO_2 nanotubes to facilitate charge separation and promote rapid charge transport under illumination [89].

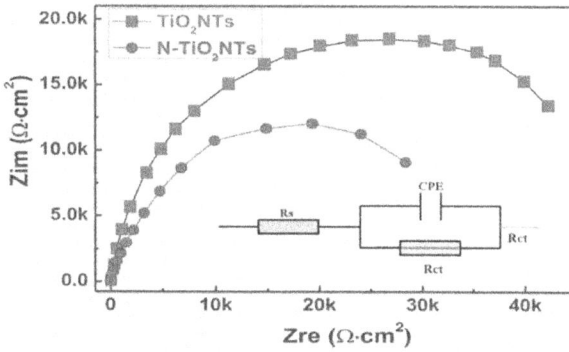

Figure 4-3: Nyquist plots of TiO$_2$ and N-TiO$_2$ nanotubes under visible light irradiation [89].

Table 4-1: Fitting results of the equivalent circuit model for TiO$_2$ and N-TiO$_2$ nanotubes [89].

Sample	R$_s$ (Ωcm^2)	R$_{ct}$ (kΩcm^2)	CPE(Fcm^{-2})
TiO$_2$	13/10	46/52	4/95
N- TiO$_2$	15/72	37/04	10/52

3- Synthesis Methods for N-Doped TiO$_2$ Nanotubes

Various synthesis techniques have been developed to fabricate nitrogen-doped TiO$_2$ nanotubes (N-TNTs), each offering distinct advantages in terms of morphology control, dopant distribution, and photoelectrochemical performance. Among these, four widely studied approaches include:

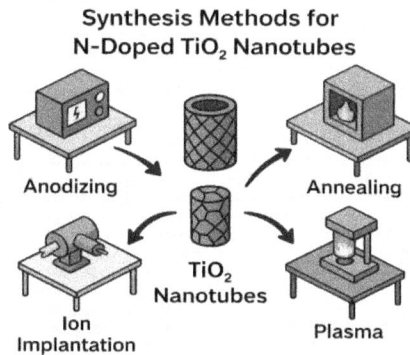

Figure 4-4: Common synthesis methods for nitrogen-doped TiO$_2$ nanotubes

3-1 Nitrogen-doped TiO$_2$ Anodizing

Nanotubes, involving the electrochemical formation of a self-organized oxide layer on a titanium substrate in the presence of oxygen-containing anions. In this process, fluoride ions (F$^-$) present in the electrolyte promote localized

33

chemical dissolution of the oxide layer, thereby aiding in the formation of TiO_2 nanotubes. The electrolysis of water at the electrolyte–metal interface generates oxygen-containing anions, which migrate toward the titanium surface, while Ti^{4+} ions simultaneously move outward from the metal. The interaction of Ti^{4+} with oxygen anions results in the formation of TiO_2, as shown in Equation (4-1). Moreover, fluoride ions are essential in facilitating both chemical dissolution and nanotube growth by reacting with Ti^{4+} ions at the oxide–electrolyte interface, as demonstrated in Equations (4-2) and (4-3) [103].

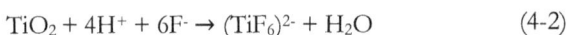

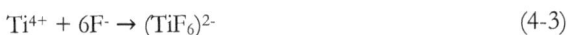

$$Ti + 2H_2O \rightarrow TiO_2 + 4H^+ + 4e^- \qquad (4\text{-}1)$$

$$TiO_2 + 4H^+ + 6F^- \rightarrow (TiF_6)^{2-} + H_2O \qquad (4\text{-}2)$$

$$Ti^{4+} + 6F^- \rightarrow (TiF_6)^{2-} \qquad (4\text{-}3)$$

In the study conducted by Li et al., nitrogen-doped TiO_2 nanotube arrays (N-TNAs) were synthesized via anodization in a mixed electrolyte composed of glycerol, water, and NH_4F. Nitrogen doping was subsequently achieved via thermal annealing at 450 °C for 3 hours, during which the nitrogen flow rate was precisely controlled within a range of 0–500 cc/min. This adjustment enabled the modulation of nitrogen content, resulting in doping concentrations ranging from 0% to 9.47% [100].

Compared to undoped TiO_2 nanotube arrays (TNAs), the N-TNAs exhibited significantly enhanced photocatalytic performance. In visible-light-driven methylene blue (MB) degradation experiments, the reaction rate constant for N-doped TiO_2 nanotube arrays (N-TNAs) reached 0.26 h⁻¹, marking an approximate 125% enhancement compared to the 0.115 h⁻¹ obtained for conventional TNAs. Moreover, recent studies have shown that employing a two-step anodization process can further improve the structural uniformity and photoelectrochemical properties of TiO_2 nanotube arrays [26].

3-2 Annealing method for N-doped TiO₂ synthesis

Thermal annealing is a widely employed post-synthesis technique for incorporating nitrogen into TiO_2 nanostructures. This process involves oxidizing Ti-based precursors at elevated temperatures in controlled atmospheres such as air, argon (Ar), or nitrogen (N_2), thereby enabling nitrogen diffusion into the crystal lattice [104].

Phuc Huo Lee et al. achieved the synthesis of nitrogen-doped TiO_2 nanotube arrays (N-TNAs) by annealing as-anodized TiO_2 structures at 450 °C for 3 hours in a pure nitrogen (N_2) atmosphere. The photocatalytic performance of both pristine and N-doped TNAs was assessed by monitoring the degradation of methylene blue (MB) under visible light illumination ($\lambda \geq$

400 nm, intensity: $120 \, mW \cdot cm^{-2}$). N-TNAs demonstrated significantly enhanced photocatalytic performance, with a reaction rate constant of $0.26 \, h^{-1}$—representing a 125% increase over the $0.115 \, h^{-1}$ value observed for conventional TNAs

This enhancement in photocatalytic performance is ascribed to the combined influence of two synergistic factors: (1) narrowing of the TiO₂ band gap induced by nitrogen-related states, which facilitates visible-light absorption, and (2) suppression of electron–hole recombination, thereby improving charge separation efficiency.

3-3 Ion Implantation for N-Doped TiO₂ Synthesis

The ion implantation method enables the direct incorporation of nitrogen atoms into oxygen lattice sites within the TiO_2 crystal structure. This modification alters the material's electronic structure by introducing localized energy states near the conduction band, thereby narrowing the band gap and enhancing photocatalytic activity under visible-light illumination .Shingang Howe and colleagues synthesized nitrogen-doped TiO_2 nanotubes (TNTs) via a combination of electrochemical anodic oxidation and nitrogen ion implantation. The synthesized materials were assessed for their photoelectrocatalytic (PEC) degradation efficiency against a range of pollutants, including methylene blue (MB), Rhodamine B, and bisphenol A (BPA). Under both UV and visible light irradiation, the N-doped TiO₂ nanotube arrays (TNTs) exhibited markedly superior PEC performance compared to their undoped counterparts. This performance enhancement is attributed to a combination of synergistic factors, including the one-dimensional tubular structure, effective nitrogen doping, and the application of external bias potential—all of which facilitate improved charge carrier separation and enhanced PEC stability. Owing to these advantages, nitrogen-doped TiO₂ nanotube arrays synthesized via ion implantation are regarded as promising materials for applications in PEC water treatment systems and dye-sensitized solar cells [105].

3-4 Plasma-Assisted Synthesis of Nitrogen-Doped TiO₂

The plasma-assisted technique offers a flexible and effective strategy for nitrogen doping of TiO_2, utilizing high-energy plasma to alter the material's electronic structure and produce reactive nitrogen species. Plasma is typically generated via direct current (DC), low-frequency discharge, or microwave discharge using radio-frequency (RF) waves. These high-energy environments facilitate the excitation of nitrogen atoms and ions, which can then interact with titanium precursors to achieve effective doping.

In this process, a titanium-based precursor solution is first vaporized and transported into the plasma reactor using an inert carrier gas. Within the reactor chamber, the precursor interacts with nitrogen species activated by the plasma, facilitating the integration of nitrogen atoms into the TiO_2 crystal lattice. This technique allows for controlled doping under low-temperature, non-equilibrium conditions, thereby preserving nanostructural integrity and minimizing unwanted phase transformations. Marcin Pesarek et al. employed a vacuum chamber integrated with a nitrogen plasma source to synthesize nitrogen-doped TiO_2 nanotubes. The setup was directly coupled with X-ray photoelectron spectroscopy (XPS) analysis, ensuring real-time monitoring of nitrogen incorporation and validating the chemical purity of the resulting material. Compared to conventional chemical and electrochemical doping strategies, the plasma method offers superior control over dopant concentration and uniformity, making it a promising alternative for producing high-performance N-TiO_2 photocatalysts [102].

.

CHAPTER 5: APPLICATIONS OF NITROGEN-DOPED TIO$_2$ NANOTUBE ARRAYS—FROM PHOTOCATALYSIS TO SUSTAINABLE ENERGY CONVERSION

1- Introduction

Nitrogen-doped TiO$_2$ nanotubes (N-TNTs) represent a versatile class of one-dimensional nanomaterials that combine the architectural advantages of ordered nanotubular arrays with the optoelectronic benefits of band-gap engineering through nitrogen incorporation. By narrowing the intrinsic band gap and introducing mid-gap states, nitrogen doping effectively extends the photoresponse of TiO$_2$ into the visible-light spectrum while simultaneously promoting charge-carrier separation and suppressing recombination [30,31,46]. Together with the inherently large surface-to-volume ratio and direct electron-transport pathways of nanotubular architectures, these modifications yield multifunctional materials capable of addressing diverse technological challenges.

As a result, N-TNTs have found broad applicability across energy and environmental domains. In photocatalysis, they have been employed for the degradation of organic pollutants, microbial disinfection, and solar-driven CO$_2$ reduction, achieving performance superior to that of conventional nanoparticulate photocatalysts [4,50,106]. In electrochemical energy storage, the enhanced conductivity and defect-driven pseudocapacitance of N-TNTs enable their use as advanced electrodes in supercapacitors, where they exhibit remarkable charge-storage capabilities and long-term cycling stability[107]. Furthermore, N-TNTs have been widely investigated as high-performance anode materials in lithium-ion batteries, where their structural robustness and improved ionic/electronic transport pathways lead to enhanced capacity retention and rate performance [85]. Beyond these, N-TNTs also play critical roles in photoelectrochemical water splitting, hydrogen evolution, fuel cell electrocatalysis, and dye-sensitized solar cells, where their tailored band structure and optimized interfacial chemistry contribute to significant improvements in energy-conversion efficiency [108–111].

This chapter provides a comprehensive overview of these applications, highlighting the unique contributions of nitrogen doping in enhancing the performance of TiO_2 nanotube arrays. By linking structure, electronic properties, and device-level outcomes, we aim to establish N-TNTs as a robust and adaptable platform for next-generation technologies in clean energy, environmental remediation, and sustainable device engineering.

2- Applications of Nitrogen-Doped TiO₂ Nanotubes

Nitrogen-doped TiO_2 nanotubes (N-TNTs) have demonstrated considerable potential across a range of applications, primarily due to their enhanced light-harvesting capability and improved charge carrier separation. The incorporation of nitrogen effectively narrows the TiO_2 band gap, enabling visible-light activation, while the tubular morphology provides a high surface area and facilitates efficient electron transport.

Owing to these favorable properties, N-TNTs have been explored for use in diverse technologies, including photocatalysis, supercapacitors, solar energy conversion, lithium-ion batteries, and water splitting. The following sections detail these applications, highlighting the specific role of nitrogen doping in enhancing functional performance.

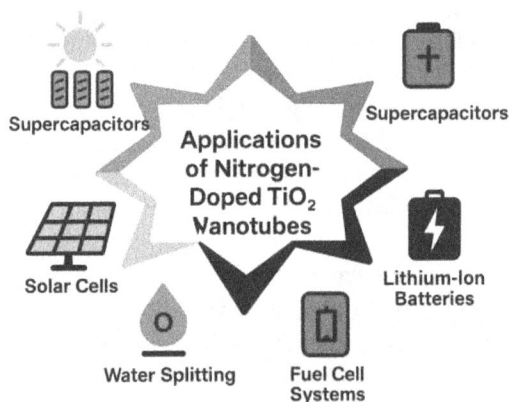

Figure 5-1. Applications of nitrogen-doped TiO_2 nanotubes in energy and environmental technologies.

2-1 Photocatalysis

The unique morphology of TiO_2 nanostructures—especially the highly ordered nanotube arrays produced through electrochemical anodization—significantly enhances light-harvesting efficiency and promotes directional charge transport, offering clear advantages over traditional nanoparticulate photocatalysts [112,113]. Owing to their structural order and efficient charge transport, TiO_2 nanotubes have found applications in various photocatalytic processes. These include degradation of organic pollutants, microbial disinfection, hydrogen evolution, and carbon dioxide reduction, in both liquid and gas-phase systems.

The photocatalytic performance of TiO_2 under both UV and visible-light irradiation is highly sensitive to its nitrogen doping level. In N-doped nanotube systems, visible-light photoactivity is significantly influenced by structural parameters such as nanotube length, wall thickness, and dopant concentration. Numerous studies have assessed the activity of N-doped TiO_2 nanotubes using organic dyes as model pollutants [100].

Mazierski et al. synthesized nitrogen-doped TiO_2 nanotubes via one-step electrochemical anodization by varying the applied voltage (20–50 V) and anodization time (30–120 min). Their results showed that both the nitrogen concentration and structural tuning played vital roles in improving key photocatalytic mechanisms, including charge carrier separation, hydroxyl radical (•OH) formation, and phenol degradation [114].

Table 5-1: Photocatalytic performance of various N-doped TiO_2-based materials under visible light.

No	Photocatalyst	Pollutant	Light Source	Degradation	Ref
1	Mo+N-TNT	MB	Visible	-	[116]
2	N-TNT	Phenol	Visible ($\lambda > 400$ nm)	96%	[114]
3	V+N-TNT		Visible	Enhanced absorption	[105]
4	Ti–N lattice N-TNT	MB, Phenol	Visible	Faster than pristine	[117]
5	Ag/N-TNT	MB	Visible	~92%	
6	Balanced N-TNT	Rhodamine B (RhB)	Visible ($\lambda > 420$ nm)	91%	[115]
7	Plasma-doped N-TNT	Methyl Orange (MO)	Visible	Enhanced (qual.)	[102]
8	N-TNT/g-C_3N_4 composite	Rhodamine B (RhB)	Visible	Strong enhancement	[118]
9	Microwave-assisted N-TNT	Methyl Orange (MO)	Fluorescent lamp (15 W)	Improved activity	[103]

Nitrogen doping alters the electronic structure of TiO_2 by introducing mid-gap states, often associated with Ti^{3+} centers and oxygen vacancies. This modification contributes to bandgap narrowing and extends the photocatalyst's activity into the visible-light spectrum. While the overall enhancement is well established, the specific contributions of interstitial versus substitutional nitrogen remain a matter of scientific debate. Some reports suggest that interstitial nitrogen leads to superior visible-light responsiveness, whereas others attribute enhanced activity to substitutional nitrogen. Interestingly, several studies indicate that a balanced coexistence of both forms yields the highest photocatalytic performance [115].

2-2 Supercapacitors

Nitrogen-doped TiO_2 nanotubes (N-TNTs) have attracted considerable interest in the field of electrochemical energy storage due to their unique combination of one-dimensional morphology, tunable surface chemistry, and stable physicochemical properties. The incorporation of nitrogen atoms into the TiO_2 lattice leads to electronic band structure modification, enhanced charge carrier density, and the creation of oxygen vacancies—factors that collectively improve the redox activity and conductivity of the electrode. As a result, N-TNTs demonstrate higher specific capacitance, better rate performance, and more stable cycling behavior compared to their undoped counterparts [107,119].

Several studies have demonstrated that electrodes based on N-TNTs exhibit superior electrochemical characteristics. Raj et al. (2018) outlined a range of TiO_2-based electrode architectures, reporting specific capacitances as high as 150 F/g with excellent retention after multiple charge–discharge cycles. Extending this baseline, Ghorbani and Khameneh Asl (2021) employed a combination of nitrogen doping and electrochemical reduction techniques, resulting in a striking capacitance of over 5666 $\mu F/cm^2$ (~2833 F/g). Their approach effectively created oxygen vacancies and optimized electron pathways, thus amplifying both the electric double-layer and pseudocapacitive behaviors crucial for high-rate performance [55,119].

Moving beyond mono-component systems, recent efforts have focused on hybridizing N-TNTs with other functional materials to amplify charge storage capability and structural integrity. Yang et al. (2020) synthesized a $TiO_2/TiN/Ti_3C_2T_x$ nanocomposite in which nitrogen doping facilitated electron mobility across the electrode matrix. Their structure achieved a specific capacitance of 361 F/g and retained 85.8% capacity after 10,000 cycles, indicating long-term cycling stability and high reversibility [120].

Another promising approach has been realized by Akbar et al. (2024), who developed a ZnSe–TiO_2 composite embedded with nitrogen-doped graphene nanosheets. The introduction of nitrogen not only enhanced electrical conductivity but also provided additional redox-active sites, resulting in a composite electrode that achieved 222 C/g with robust stability. These findings highlight the growing potential of multi-phase, nitrogen-enriched nanostructures in achieving high-rate capabilities, especially when paired with active transition metal components [121].

In summary, the strategic implementation of nitrogen-doped TiO_2 nanotubes—either as standalone electrodes or within hybrid architectures—has shown remarkable promise for next-generation supercapacitor systems. The use of dopants such as nitrogen introduces multifunctional advantages, including increased conductivity, defect-driven pseudocapacitance, and enhanced interfacial charge transport. Continued research into the precise control of doping mechanisms and the integration of complementary materials will be critical to unlocking the full potential of N-TNT-based supercapacitors [107,121].

Table 5-2: Summary of studies on nitrogen-doped TiO_2 nanotubes as supercapacitor electrodes.

Title	Authors Year	Material/Method	Performance	Key Finding	Ref
Review—Advent of TiO_2 Nanotubes as Supercapacitor Electrode	Raj & Prasanth 2018	Review on TiO_2 NTs	Up to 150 F/g	Summarized applications and performance of TiO_2 NTs	[55]
Electrochemical Performance of Nitrogen-Doped TiO_2 Nanotubes	Appadurai et al. 2019	N-doped TiO_2 nanotubes via anodization	~ 835 $\mu F/cm^2$ (~ 417 F/g)	Improved conductivity and double-layer capacitance	[107]
Nitrogen Doped Intercalation $TiO_2/TiN/Ti_3C_2T_x$ Nanocomposite Electrodes	Zhang et al. 2020	$TiO_2/TiN/$ MXene nanocomposite with N-doping	361 F/g at 1 A/g	High capacity and cycling stability	[120]
Modifying TiO_2 nanotube using N-doping and electrochemical reductive doping	Ghorbani & Khameneh Asl 2021	N-doping + electrochemical reduction	~ 5666 $\mu F/cm^2$ (~ 2833 F/g)	High pseudocapacitance due to oxygen vacancies	[119]
Exploring the Potential of Nitrogen-Doped Graphene in ZnSe–TiO_2 Composite Materials	Akbar et al. 2024	ZnSe–TiO_2 with N-doped graphene	222 C/g	Enhanced conductivity and redox-active surface	[121]

2-3 Lithium-Ion Batteries

Ongoing progress in lithium-ion battery (LIB) technology has established these systems as the leading choice for energy storage in diverse applications, ranging from portable electronic devices to electric and hybrid vehicles. Their broad utilization stems from advantages such as high energy density, low self-discharge rates, improved safety mechanisms, and overall operational efficiency. LIBs are also key components in renewable energy storage, enabling effective energy retention from sources such as solar and wind power systems. Despite these advantages, there remains an increasing demand to further enhance the capacity, stability, and long-term performance of LIBs to meet the evolving needs of modern electronics. While theoretical improvements can be achieved using pure lithium metal, this approach often introduces challenges related to electrolyte compatibility and voltage instability. In response to these challenges, various alternative electrode materials have been investigated. Among them, titanium dioxide (TiO_2) has emerged as a promising candidate to replace conventional graphite anodes, owing to its superior safety characteristics, low toxicity, structural robustness, and cost-effectiveness. Although TiO_2 exhibits limited electrical conductivity, its physicochemical characteristics and tunable nanostructures—such as nanotubes—offer promising pathways to improve lithium-ion transport, mitigate volume expansion, and enhance overall electrochemical performance [123].

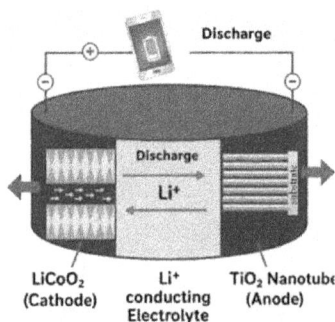

Figure 5-2: Schematic representation of the fundamental working principle of Li^+ batteries [123].

The application of anatase-phase TiO_2 as an anode material in lithium-ion batteries has been extensively investigated, with particular emphasis on its practical viability. One of the key challenges limiting its capacity is the structural phase transition from a tetragonal to an orthorhombic configuration during lithiation, which affects structural stability and electrochemical performance. Additionally, the intrinsically low electronic and ionic conductivity of TiO_2 restricts its application in high-performance energy storage systems [124].

To overcome these limitations, researchers have explored several approaches. One strategy involves engineering TiO_2 at the nanoscale to increase surface area and shorten lithium-ion diffusion pathways. Nanostructured TiO_2, especially in forms like nanotubes or nanorods, exhibits higher surface-to-volume ratios, which enhances lithium intercalation kinetics and reduces charge transfer resistance [125].

Furthermore, doping TiO_2 with various elements—both metals and non-metals—has shown promising effects in improving conductivity and storage capacity. Among non-metals, nitrogen doping has received particular attention. Its ability to introduce donor states and alter the band structure makes N-doped TiO_2 a viable candidate for improving lithium-ion transport and enhancing the overall charge/discharge behavior of the electrode. The thickness and uniformity of nitrogen incorporation are especially critical factors in achieving optimal battery performance [126].

Recent experimental studies have demonstrated the practical advantages of nitrogen doping when combined with nanostructuring and hybrid material desig. To address the inherent limitations of pristine TiO_2 as an anode material, recent research has focused extensively on strategic doping and structural hybridization. Nitrogen doping, in particular, has emerged as a powerful approach to enhance the electrical conductivity and lithium-ion diffusion capability of TiO_2-based systems. When TiO_2 is doped with nitrogen, it introduces donor states and oxygen vacancies, modifying the electronic structure and reducing the charge transfer resistance at the electrode–electrolyte interface.

For example, Li et al. (2019) successfully engineered a nitrogen-doped TiO_2/reduced graphene oxide (rGO) hybrid anode. This composite delivered an initial specific capacity of 318 mAh/g and maintained 210 mAh/g after 2000 charge–discharge cycles. The superior performance was attributed to the formation of Ne–Ti–O and Ne–C bonds that promoted both structural cohesion and rapid lithium-ion transport across the hybrid framework. Their findings highlight the synergistic effect of nitrogen doping and carbon support in achieving prolonged electrochemical stability [127].

In another prominent study, Han et al. (2017) developed few-layer TiO_2–B nanosheets integrated with nitrogen-doped graphene. This advanced structure exhibited an impressive initial capacity of 583.3 mAh/g and retained 326.4 mAh/g after 100 cycles. Even under extreme current conditions (50C), the material preserved 153.3 mAh/g, showcasing its exceptional rate capability. The integration of a layered TiO_2–B matrix with conductive, nitrogen-rich graphene allowed efficient electron mobility and minimized structural degradation over extended cycling [128].

Mechanistic insights were provided by Appadurai et al. (2017), who employed in situ transmission electron microscopy (TEM) to investigate the lithiation behavior of N-doped anatase TiO_2 nanotubes. The real-time observations confirmed a three-stage lithiation process and demonstrated how nitrogen atoms embedded within the TiO_2 lattice helped maintain crystallographic coherence during cycling. This study provided direct evidence of the structural benefits imparted by nitrogen doping, further validating its critical role in achieving long-lasting anode materials [129].

Table 5-3: Summary of studies on nitrogen-doped TiO$_2$ nanotubes as anode materials in lithium-ion batteries.

Material/System	Synthesis Method	Performance (mAh/g)	Cycle Stability	Key Finding	ref
N-doped TiO$_2$/rGO	Hydrothermal + thiourea calcination	318 initial, 210 after 2000 cycles	~66% retention	Ne–Ti–O and Ne–C bonds improved conductivity and Li$^+$ diffusion	[127]
N-doped TiO$_2$–B/Graphene	Hydrothermal + N-doping	583.3 initial, 326.4 after 100 cycles	~67.8% retention	Improved high-rate capability and structural robustness	[128]
N-doped anatase TiO$_2$ nanotubes	Anodization + NH$_3$ treatment	~200 @0.1C to ~100 @10C	Stable at high rate	Low R$_{ct}$ and oxygen vacancies enhanced Li-ion intercalation	[129]
N-TiO$_2$ nanotubes	Anodization + NH$_3$ cracking	975 μAh/cm^2 (initial), 145 after 200 cycles	>98% efficiency	High areal capacity and excellent rate capability	[107]
N-doped TiO$_2$ nanotubes	Thermal oxidation + NH$_3$ doping	Enhanced capacity and cycling	Improved rate retention	Clear role of N sites in improving electron/ion transport	[130]
TiO$_2$/N-doped carbon on Si	Recycling-based carbonization + TiO$_2$	Enhanced vs. pristine Si	High stability over 200+ cycles	Eco-friendly hybrid improves performance of Si anodes	[131]
TiO$_2$/N-doped CNF	Electrospinning + N-rich precursor	High capacity at low rate	Good capacity retention	Synergistic effect of TiO$_2$ and conductive CNF matrix	[132]
Various TiO$_2$-based systems	Multiple synthesis approaches	Varied	Varied	Summarized advances in TiO$_2$ nanostructures for LIBs	[133]
Mn/F/N co-doped TiO$_2$(B)	Co-doping strategy + annealing	Stable under high current	Good retention >200 cycles	Multi-element doping improves capacity in harsh conditions	[134]

2-4 Application of TiO$_2$ Nanotubes in Fuel Cell Systems

Owing to their semiconducting properties and large specific surface area, titanium dioxide (TiO$_2$) nanotubes have recently attracted growing interest in the field of fuel cell research. In 2011, Hoseini et al.[135] synthesized TiO$_2$

nanotubes via anodic oxidation and employed gold nanoparticles—electrochemically deposited from a solution containing $KAu(CN)_2$ the surface of these nanotubes. These gold nanoparticles acted as catalysts for the oxidation of hydrazine, a promising fuel candidate for fuel cell applications.

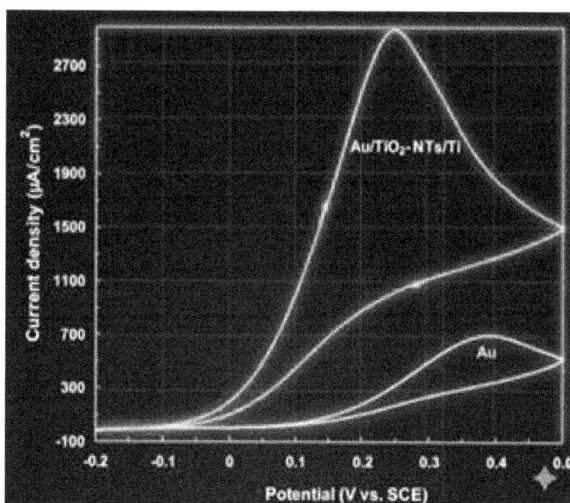

Figure 5-3: buffer solution containing 0.00085 M hydrazine at 25°C, recorded at a scan rate of 100mV/s [135].

To evaluate the performance, cyclic voltammetry was employed. The findings revealed that in a phosphate buffer solution containing hydrazine at 25 °C, the TiO₂ nanotube electrode modified with gold nanoparticles exhibited a markedly higher current density compared to that of the pure gold electrode. This enhancement in electrochemical activity is illustrated in Figure 5-3.

In another study conducted in 2012, Hoseini et al. [136] investigated the electrocatalytic performance of a titanium dioxide electrode decorated with platinum nanoparticles for the oxidation of galactose. The results demonstrated that the TiO₂–Pt composite electrode delivered a significantly higher current density than both the unmodified TiO₂ nanotube electrode and the pure platinum electrode, under both alkaline and acidic conditions. As shown in Figure 5-4, this improvement is mainly attributed to the increased specific surface area of the modified electrode, which provides a greater number of electroactive sites compared to a flat platinum surface.

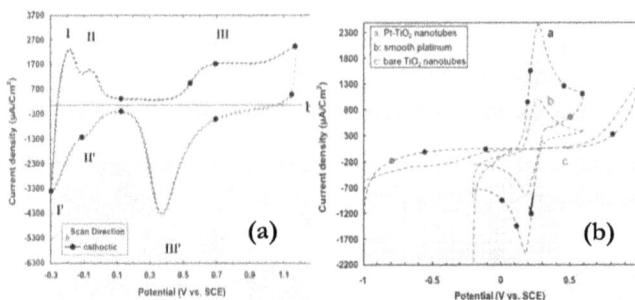

Figure 5-4: (a) CV curves of Pt–TiO$_2$ and bare TiO$_2$ nanotube electrodes in 0.1 M H$_2$SO$_4$ at 100 mV/s; inset shows flat Pt electrode response. (b) CV responses of Pt–TiO$_2$, pure Pt, and bare TiO$_2$ in 1 M KCl + 10 mM Fe(CN)$_6$ at 25°C, scan rate: 100 mV/s [136].

2-5 Water Splitting

The growing global concern over energy demand and climate change has intensified efforts to identify clean and renewable energy alternatives. Among these, hydrogen has emerged as a promising sustainable fuel, owing to its high energy density and zero emission of harmful by-products during utilization. Hydrogen generation through photoelectrochemical (PEC) water splitting not only offers a renewable path for energy production but also enables the simultaneous degradation of organic pollutants via oxidation reactions [12].

In a typical PEC process, water undergoes redox reactions: hydrogen ions are reduced to form H$_2$ gas, while water molecules are oxidized to generate O$_2$. For effective hydrogen generation, the semiconductor's conduction band (CB) must be positioned at a potential more negative than the hydrogen evolution potential, while the valence band (VB) should lie at a potential more positive than 1.23 V versus the normal hydrogen electrode (NHE) to facilitate oxygen evolution [108].

Among various semiconductor materials explored for PEC water splitting, TiO$_2$ remains a strong candidate due to its favorable band edge positions, chemical stability, environmental benignity, and cost-effectiveness (Figure 5-5a) [137]. Upon light irradiation (Figure 5-5b), photoexcited electrons are transferred from the valence band (VB) to the conduction band (CB) of TiO$_2$. These electrons contribute to the reduction of protons (H$^+$) into hydrogen gas (H$_2$), while the photogenerated holes in the VB facilitate the oxidation of water molecules, leading to oxygen (O$_2$) evolution. This process can also contribute to the breakdown of toxic organic compounds when present in solution.

Although conventional photocatalysis faces limitations due to rapid electron–hole recombination, PEC systems help overcome this issue by applying an external bias and using a built-in electric field to separate charge carriers more effectively. As a result, PEC water splitting offers a highly efficient route for hydrogen production and environmental remediation [138].

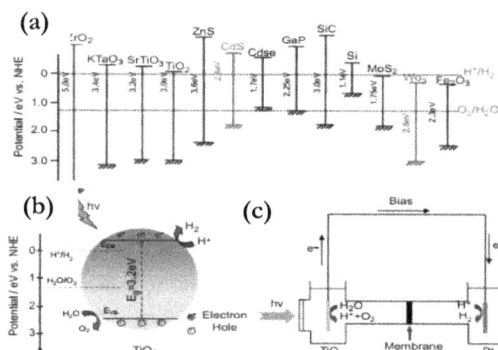

Figure 5-5: (a) Band structure vs. redox potentials for water splitting; (b) photocatalytic and (c) photoelectrocatalytic water splitting on TiO_2 [138].

As illustrated in Figure 5-5c, applying a low bias potential to TiO_2 nanotube arrays (TNTs) significantly improves their photoelectrochemical performance. The application of an external potential suppresses the recombination of photogenerated electron–hole pairs and facilitates charge carrier transport across the electrode–electrolyte interface. Under strong ultraviolet (UV) illumination, electrons are excited from the valence band (VB) to the conduction band (CB) of TiO_2 and are efficiently directed toward the external circuit. The applied bias drives these electrons to the counter electrode, while the remaining holes in the VB initiate oxidation reactions at the photoanode surface. This effective charge separation leads to the generation of reactive species: at the cathode, electrons reduce protons to produce hydrogen gas, whereas at the anode, holes oxidize water to form oxygen or degrade organic pollutants.These phenomena are consistent with the fundamental principles of photocatalysis and photoelectrochemical hydrogen production [139].

$$TiO_2 + hv \rightarrow e\text{-} + h^+ \tag{5-1}$$

$$2H^+ + 2e^- \rightarrow H_2 \tag{5-2}$$

$$2H_2O + 4h^+ \rightarrow O_2 + 4H^+ \tag{5-3}$$

Overall reaction $\qquad : 2H_2O \rightarrow O_2 + 2H_2 \tag{5-4}$

TiO_2 nanotubes have gained widespread application in water treatment and solar-driven hydrogen production due to their well-ordered tubular structure, high surface area for particle exchange, and relatively long lifetimes for photogenerated electron–hole pairs. These structural and electronic features make them effective platforms for charge transport and photocatalytic reactions. However, despite their potential, several challenges still limit the practical use of TiO_2 nanotubes in solar water splitting. These limitations involve rapid recombination of photogenerated charge carriers under certain conditions, inadequate absorption of visible light, and suboptimal utilization of the entire solar spectrum. Additionally, structural or interfacial losses may hinder efficient charge transfer. As a result, extensive research continues to focus on overcoming these limitations by modifying the material's electronic properties, surface chemistry, and light-harvesting capability [138].

2-6 Solar Cell

The development of efficient dye-sensitized solar cells (DSSCs) has garnered considerable interest in recent years, particularly through the modification of semiconductor photoanodes. Among these, titanium dioxide (TiO_2) nanotubes have emerged as a promising material due to their one-dimensional electron transport path and high surface area. A crucial strategy to enhance their performance is nitrogen doping, which introduces mid-gap states and narrows the bandgap, thereby improving visible light absorption and charge carrier separation [140].

Nitrogen-doped TiO_2 nanotubes (N-TiO_2) have demonstrated significant improvements in power conversion efficiency (PCE) by enhancing the photoresponse under solar irradiation. For instance, introducing N atoms into the TiO_2 lattice through annealing in ammonia environments or urea decomposition alters the electronic structure and boosts photocurrent generation [141]. In comparative studies, N-doped nanotube arrays exhibited higher Jsc and Voc values than undoped counterparts, highlighting the role of doping in tuning electronic properties for optimal photovoltaic performance [142].

Moreover, device architecture and the substrate interface significantly influence the doping effect. TiO_2 nanotube electrodes fabricated on titanium foils and subsequently doped with nitrogen have shown improved electron lifetime and reduced charge recombination due to enhanced interfacial bonding and electron transport efficiency [142]. A comprehensive review on TiO_2 nanotube-based DSSCs further affirms that nitrogen doping is among the most effective strategies for extending photoresponse into the visible range while retaining chemical stability [143].

Beyond efficiency improvements, the mechanism of nitrogen incorporation plays a pivotal role in defining device behavior. Studies on solid-state DSSCs suggest that localized states introduced by N atoms may facilitate or hinder charge transfer depending on the doping concentration and annealing conditions. These findings emphasize the need for precise control over doping profiles during fabrication [144].

In addition, the application of nitrogen-doped nanotubes in DSSCs is not limited to single-dopant systems. Some research has explored co-doping strategies, combining nitrogen with other nonmetals or plasmonic nanoparticles, to further synergize light absorption and carrier dynamics [145].

Table 5-4: Summary of studies on nitrogen-doped TiO$_2$ nanotubes performance of TiO$_2$-based photoanodes in DSSCs.

Title	Focus	Material Form	Doping Method	Efficiency Impact	Ref
Effect of N Dopant Amount on the Performance of DSSCs Based on N-Doped TiO$_2$ Electrodes	Impact of N-doping concentration on DSSC performance	Nanotube Array	Annealing in NH$_3$	Increased efficiency ~20%	[140]
Nitrogen-doped TiO$_2$ nanoparticles: Better TiO$_2$ nanotube array photo-anodes for DSSCs	Enhancement of photoanode performance with N-doped TiO$_2$	Nanoparticles	Sol-gel with NH$_4$NO$_3$	Enhanced Jsc and Voc	[146]
Dye-sensitized Solar Cell Based on N-Doped TiO$_2$ Electrodes Prepared on Titanium	TiO$_2$ nanotubes on Ti substrate with N doping for DSSC	NT on Ti foil	Anodization + N doping	Moderate increase	[147]
TiO$_2$ Nanotubes for Dye-Sensitized Solar Cells — A Review	Comprehensive review of TiO$_2$ nanotube applications in DSSCs	Various (Review)	Multiple	Summarized trends	[143]
Influence of Nitrogen Doping on Device Operation for TiO$_2$-Based Solid-State DSSCs	Photo-physics and device operation influenced by N doping	Solid-state device	Plasma doping	Improved charge transport	[144]
Application of Nitrogen-Doped TiO$_2$ Nano-Tubes in Dye-Sensitized Solar Cells	N-doped TiO$_2$ nanotubes in DSSCs to improve efficiency	Nanotubes	Annealing	Enhanced η from 4.7% to 7.3%	[145]
Synthesis and Characterization of Various Doped TiO$_2$ Nanocrystals for DSSCs	Comparative study of doped TiO$_2$ nanocrystals in DSSCs	Nanocrystals	Wet chemical synthesis	Comparative performance	[148]
Optimization of N Doping in TiO$_2$ Nanotubes for Enhanced Solar Light Mediated H$_2$ Production and Dye Degradation	Optimizing N doping for photocatalysis and solar H$_2$	Nanotubes	Urea-assisted doping	Enhanced H$_2$ yield + DSSC potential	[88]

The synthesis of nanocrystals with varied doping schemes provides a comparative platform for identifying optimal formulations tailored to specific light environments or dye molecules [6].

Recent advancements also indicate that the benefits of N-doped TiO_2 extend beyond photovoltaic efficiency. Optimized doping configurations have demonstrated dual functionalities, supporting both solar-driven hydrogen production and organic dye degradation, suggesting potential in hybrid photoelectrochemical devices [88].

In summary, nitrogen doping of TiO_2 nanotubes remains a robust and versatile method for enhancing the performance of dye-sensitized solar cells. Ongoing research continues to refine doping techniques, integrate multifunctionality, and unlock new application potentials.

3- Fundamentals of Photoelectrocatalysis

Both photocatalysis (PC) and photoelectrocatalysis (PEC) operate on the same fundamental concept: when a semiconductor absorbs light with energy equal to or greater than its band gap, an electron from the valence band (VB) gets excited into the conduction band (CB), leaving behind a positively charged hole in the VB. This process creates an electron–hole pair (e^-/h^+). These charge carriers can drive chemical reactions, and their reactivity is determined by the energy positions of the CB and VB, as illustrated in Figure 5-6a. In TiO_2-based photocatalytic systems, the photogenerated holes (h^+) have strong oxidative power. They can react with water molecules to generate hydroxyl radicals (•OH), which are highly reactive species (Equation 5-6). At the same time, the excited electrons in the conduction band (CB) can reduce oxygen (O_2) to form superoxide radicals ($O_2\bullet^-$) (Equation 5-7). One key drawback of conventional photocatalysis is the rapid recombination of these electrons and holes. This recombination limits the number of charge carriers available for redox reactions, thereby reducing the overall efficiency of the process [108,149,150].

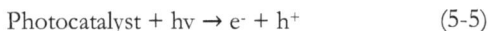

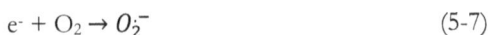

$$\text{Photocatalyst} + h\nu \rightarrow e^- + h^+ \qquad (5\text{-}5)$$

$$h^+ + H_2O \rightarrow HO^\bullet_{ad} + H^+ \qquad (5\text{-}6)$$

$$e^- + O_2 \rightarrow O_2^{\bullet-} \qquad (5\text{-}7)$$

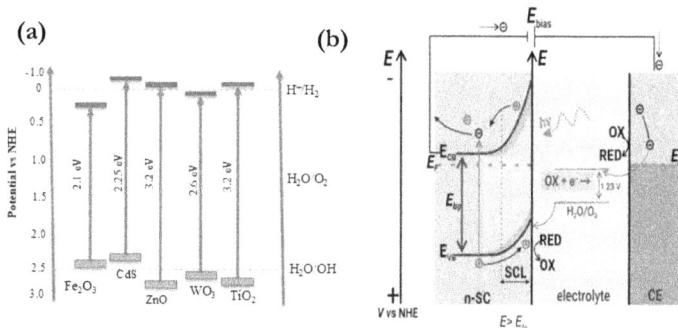

Figure 5-6: (a) Band edge positions of common semiconductors at pH 0; (b) schematic of an n-type PEC cell under illumination and applied bias [150].

Photoelectrochemical (PEC) systems offer a more efficient solution than traditional photocatalysis (PC) by enhancing both light absorption and charge carrier separation [151]. In a typical PEC setup, the photocatalyst is coated on a conductive substrate, forming a photoanode that operates under an externally applied electric field [152,153]. This configuration improves the separation and direction of electrons and holes, which significantly increases the degradation rate of pollutants. As shown in Figure 5-6b, PEC systems generally achieve better performance than conventional PC or electrocatalytic methods in removing contaminants [154,155].

In n-type semiconductors, applying a positive bias voltage (greater than the flat-band potential) helps enhance charge separation. This applied potential shifts the Fermi level and causes band bending, which pushes electrons and holes in opposite directions—toward the cathode and anode, respectively [156]. As a result, holes accumulate at the surface of the photoanode, where they drive oxidation reactions. Meanwhile, electrons are efficiently extracted from the material's bulk and directed toward the counter electrode. This separation mechanism greatly reduces recombination losses and significantly boosts the performance of PEC systems.

3-1 Mechanism of photoelectrocatalyst

With rapid economic development, environmental issues—particularly the scarcity of clean water and increasing levels of water pollution—have become increasingly critical. There is an urgent global demand for effective solutions to address these challenges. In this context, titanium dioxide (TiO₂) has gained recognition as an exceptional photocatalyst owing to its affordability, non-toxic nature, chemical stability, and strong oxidative power, which allows it to effectively degrade various hazardous organic pollutants [157].

In photoelectrochemical applications, a TiO₂ film electrode—referred to as a photoanode—is commonly employed as the working electrode. An appropriate bias potential is applied to this electrode to facilitate charge separation and carrier transport [158]. This applied bias drives photogenerated electrons from the TiO_2 photoanode through an external circuit, directing them toward the counter electrode. Consequently, reduction reactions occur at the counter electrode, whereas oxidation processes are facilitated at the surface of the TiO_2-based photoanode [153].

3-2 Factors Affecting the Efficiency of the Photoelectrocatalyst Process

The efficiency of a photoelectrocatalytic (PEC) system depends on a combination of interrelated physical, chemical, and operational parameters. As shown in the figure 5-7, key parameters—such as photoanode type and mobility, light source features, counter electrode composition, film thickness, and electrolyte characteristics (e.g., pH, conductivity)—critically affect PEC performance. Each of these elements influences key aspects such as charge generation, carrier separation, interfacial reactions, and redox kinetics. Understanding and optimizing these variables is essential for designing effective PEC systems for environmental and energy-related applications.

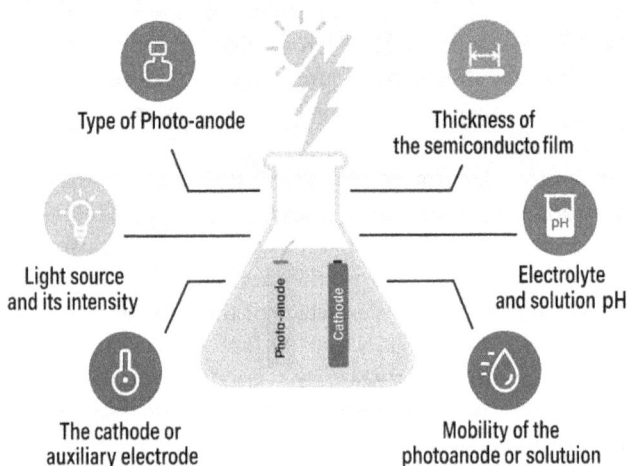

Figure 5-7: Key factors influencing photoanode performance in photoelectrochemical systems.

3-2-1 Photoanode type

Common photocatalysts are typically constructed using titanium (Ti) meshes or substrates coated with TiO_2 [159–161], as well as $Ti/Ru_{0.3}Ti_{0.7}O_2$ composite materials [159,162–166]. Ti/TiO_2 electrodes are generally preferred in three-electrode systems under potentiostatic conditions due to their stability at low current densities. In contrast, $Ti/Ru_{0.3}Ti_{0.7}O_2$ demonstrates excellent durability at higher current densities, making it suitable for application in two-electrode reactor configurations. This composite structure facilitates more efficient dye degradation by enabling current density adjustments while maintaining structural integrity.

Janaki et al. [167] investigated carbon-based nanostructured electrodes as substrates for the electrochemical deposition of metal and semiconductor oxides. Their findings emphasized the benefits of these materials in applications such as photoelectrochemical solar cells and heterogeneous photocatalysis, owing to their excellent electrical conductivity and advantageous physicochemical characteristics.

Extensive efforts have also been devoted to engineering TiO_2-based photocatalysts with enhanced specific surface area. Among various synthetic approaches, the use of TiO_2 powder remains one of the most prominent. In particular, Degussa P25—produced through the hydrolysis of $TiCl_4$—is widely recognized as a benchmark commercial photocatalyst. Despite its popularity, the use of TiO_2 powder presents two main challenges: limited visibility due to its fine particulate nature and difficulties in recovery and recycling [168,169]. These limitations can be mitigated by combining multiple strategies aimed at enhancing both environmental compatibility and catalyst reusability.

3-2-2 Light Source and Intensity

In numerous studies, the selection of a suitable light source for photoelectrochemical (PEC) applications is primarily determined by the band gap and corresponding absorption wavelength of the photocatalyst used [170–172]. Based on this consideration, three major illumination categories are commonly employed: UVC ($\lambda < 300$ nm), UVA ($\lambda = 320$–400 nm), and visible light (UV/Vis), which includes both artificial and natural sunlight. UV light is widely used due to its high photon energy and strong ability to excite electrons in semiconductors such as TiO_2 and ZnO.

In various experimental configurations, artificial UVC light has been utilized to activate photocatalysts, resulting in efficient degradation of organic contaminants. Conversely, visible light—which accounts for approximately 43% of the solar spectrum reaching Earth's surface—is generally less effective for many conventional photocatalysts due to their inherently wide band gaps. This limitation has driven substantial research aimed at modifying photocatalyst materials to narrow their band gaps and extend their light absorption capabilities into the visible region. In this context, semiconductors like $BiVO_4$ and Fe_2O_3, which are responsive to visible light, have demonstrated significant potential [173].

As detailed in Table 5-5, light intensity also plays a critical role in photocatalytic activity. An increase in light intensity generally necessitates a proportional rise in current density to maintain efficient transport of photoexcited electrons from the photocatalyst's conduction band to the cathode [170,172]. Numerous studies have examined the influence of light intensity on the oxidative degradation of organic pollutants, consistently demonstrating a positive correlation between irradiation level and degradation efficiency [174,175]. Exposure to UV light not only enhances photocatalytic degradation rates but also accelerates the inactivation of bacteria and organic compounds. This enhancement is attributed to the generation of more h_VB^+ holes and reactive oxidizing species, particularly hydroxyl radicals (•OH), on the photocatalyst surface [172]. However, efficient electron transfer to the cathode under such illumination conditions demands a higher current density across the system to maintain charge balance and enhance overall PEC performance [176].

Table 5-5: Semiconductor photocatalysts are widely used in PEC photoanodes.

Photocatalayst	Crystaline phase	Band gap (eV)	Adsorption band (nm)
TiO₂	Rutile	3.05	405
	Anatase	3.25	385
WO₃	Tungstite	2.5–2.7	500
ZnO	Wurtzite	3.2–3.4	~400
SnO₂	Cassiterite	3.6	~345
α-Fe₂O₃	Hematite	2.2	Visible-light
BiVO₄	-	2.4 – 2.5 eV	Visible-light

3-2-3 Selection of Cathode or Auxiliary Electrode

The effectiveness of conventional PC systems is often undermined by the swift recombination of photogenerated electrons and holes[172,175,177]. In photoelectrochemical (PEC) systems, the issue of charge carrier recombination can be effectively addressed by optimally selecting a cathode or auxiliary electrode that functions as an efficient electron acceptor. This electrode facilitates the extraction of electrons from the conduction band, thereby suppressing recombination processes and significantly enhancing the system's overall oxidation efficiency [178]. The overall configuration of a PEC system is designed to maintain electrochemical equilibrium, with reduction reactions occurring at the cathode and oxidation at the photoanode.

One of the most valuable advantages of PEC systems in wastewater treatment is their ability to produce hydrogen as a by-product of water electrolysis—a process that currently accounts for only a small fraction of global hydrogen production [179,180]. A broad range of materials have been employed as cathodes in photoelectrochemical (PEC) systems, including platinum [181–183], titanium, copper [184], stainless steel [185,186], carbon nanotubes, dimensionally stable anodes (DSAs) [187], graphite [188,189], boron-doped diamond, and dimensionally stable anodes [172,178,190]. Some carbon-based cathodes—such as graphite, glassy carbon, and carbon felt—have also shown potential in generating electrochemical hydrogen peroxide (H₂O₂). Notably, oxygen-enhancing cathodes and boron-doped diamond electrodes have demonstrated this capability as well.

The physical configuration of the cathode can vary based on the specific design of the photoelectrochemical (PEC) reactor. Common cathode geometries include flat plates, discs, hollow cylinders, meshes, reservoir-type structures, and transparent conductive materials such as indium tin oxide (ITO) glass [179]. Additionally, enzyme-assisted cathodes have demonstrated the potential to enhance system performance by improving energy conversion efficiency and lowering oxygen demand, thereby boosting H_2O_2 production. For example, Xia et al. [191] achieved a production yield of 5.4 mmol H_2O_2 under optimized conditions using a dopamine-modified carbon felt cathode (CF-DPA). In a similar approach, Cheng et al. [192] utilized hemin/copper-functionalized carbon fibers as cathodes to facilitate efficient enzyme-assisted and electrochemical generation of H_2O_2. In another example, Chen et al. [193] demonstrated that customizing the cathode design for specific applications—such as iron-EDTA oxidation and iron recovery—can significantly enhance performance. Their study showed that employing a rotating cathode greatly increased the generation of metal ions in contact with the electrode surface, thereby improving the overall metal recovery efficiency.

3-2-4 Thickness of the Semiconductor Film on the Electrode Surface

The thickness of the semiconductor film on the electrode surface plays a crucial role in determining the photocatalytic efficiency of TiO_2-based systems, especially in dual-layer configurations. Precise control over film thickness is essential to optimize light absorption, charge separation, and interfacial electron transfer processes. The quality and quantity of the TiO_2 layer significantly affect UV light absorption and its interaction with reactive sites. Both insufficient and excessive film thickness can hinder performance—thin layers may lead to inadequate photon absorption, while overly thick films can cause light scattering or reflection, reducing the number of photons reaching the catalytic interface.

Although studies directly linking reaction rate to TiO_2 film thickness are limited, it is well understood that optimal thickness ensures efficient light penetration and electron–hole generation at catalytically active sites. For example, in configurations where a glass electrode is first modified with a SnO_2 layer and subsequently coated with TiO_2, the maximum optical penetration depth is governed by the parameter $1/\alpha$, where α represents the absorption coefficient of TiO_2 at the specific irradiation wavelength (Figure 5-8).

Electron–hole pairs generated within the region spanning from W to (W − Lp), where W denotes the depletion width and Lp represents the minority carrier diffusion length (approximately 100 nm for TiO_2)—can be efficiently transported to the electrode surface prior to recombination [194]. Beyond this active region, photogenerated carriers are more likely to recombine before contributing to redox reactions. Additionally, the distribution of carriers toward the depletion region and the surface is influenced by the internal electric field, which is in turn affected by doping levels and the presence of surface-bound negative species (Figure 5-8) [194,195].

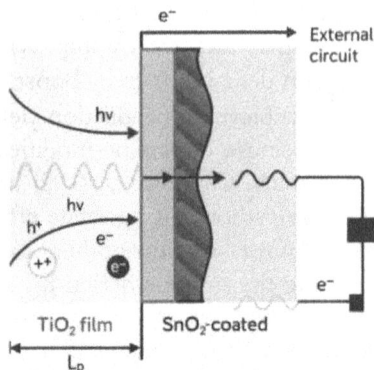

Figure 5-8:: illustrates the patterning of n-type TiO_2 on SnO_2-coated glass under ultraband light, highlighting the light penetration depth $(1/\alpha)$, depletion width (W), and minority carrier diffusion length (Lp) [175].

3-2-5 Electrolyte and Solution PH

The electrolyte plays a critical role in photoelectrochemical (PEC) systems, as it facilitates the completion of both electrolysis and oxidation reactions. By increasing the solution's ionic conductivity, the electrolyte helps reduce ohmic voltage drop and lowers the overall electrical energy consumption during operation [172,176,196].

Moreover, the electrolyte composition significantly influences the kinetics of pollutant degradation, acting as a mediator in the generation of reactive species. In saline wastewater, for example, the presence of high concentrations of sodium chloride (NaCl) leads to the formation of reactive chlorine species, which are highly effective in degrading organic contaminants [191,197,198].

The pH of the solution is another critical parameter that significantly influences photocatalytic efficiency. It not only influences the formation and behavior of reactive intermediates, such as hydroxyl radicals and superoxides, but also determines the charge state of the photocatalyst surface, thereby controlling its ability to adsorb target pollutants [190,199,200]. Additionally, the pH directly impacts the stability of the photoelectrode and the corrosion kinetics of materials used in PEC systems.

This pH dependency was clearly illustrated in the degradation of 4-chlorophenol using a TiO_2 nanotube photoanode operated under a constant bias of 0.6 V and an illumination intensity of 2.5 mW·cm⁻². Results showed that acidic conditions yielded significantly higher degradation rates compared to alkaline environments.

Other operational parameters also influence the performance of photoelectrochemical (PEC) systems. For example, dissolved oxygen (O_2) plays a key role in suppressing the recombination of photogenerated electron–hole pairs, thereby facilitating the formation of reactive oxygen species such as superoxide radicals. Similarly, temperature changes influence the viscosity of the solution and mass transfer rates. While elevated temperatures can enhance mass transport toward the anode, they may also alter the reaction kinetics and degradation pathways of pollutants [201].

Stirring speed is another critical operational parameter, as it governs the rate of mass transfer and promotes the homogeneous distribution of reactants within the reactor. Moreover, the ratio of effective electrode surface area to solution volume, along with the pollutant concentration, plays a decisive role in determining the overall efficiency of organic pollutant degradation in aqueous systems [175–177]. These interrelated factors must be carefully optimized to achieve maximum PEC system performance.

3-2-6 Mobility of the Photoanode or Solution

The quantum efficiency and direct oxidation capability in photoelectrochemical (PEC) systems are closely linked to the intensity and path of incident light. In typical PEC reactor designs, the photoanode is fully immersed in the electrolyte solution, meaning that incoming light must first pass through the reactor wall and the solution medium before reaching the surface of the photoanode [202–204]. This setup often leads to energy losses due to light scattering and absorption within the solution, reducing the effective light intensity at the semiconductor surface and, consequently, decreasing photocatalytic efficiency.

Figure 5-9 illustrates the electrochemical response of 1×10^{-4} mol·L^{-1} p-nitrophenol (pNP) in 0.1 mol·L^{-1} H$_2$SO$_4$, recorded by differential pulse voltammetry (DPV) using a platinum disk electrode modified with nano-TiO$_2$. Under dark conditions, the current in the potential window of +0.40 to +0.70 V remained very low and nearly constant, indicating that the TiO$_2$ surface was not photoactive and that no significant electrochemical oxidation of pNP occurred in the absence of irradiation. Upon exposure to UV light, however, a progressive anodic response emerged after approximately 15 minutes, reflecting the photoactivation of TiO$_2$ via electron–hole pair generation and the subsequent production of hydroxyl radicals (•OH).

The most prominent anodic peak appeared at approximately +0.55 V, which can be assigned to the oxidation of intermediate species generated by the interaction of •OH with pNP molecules. This observation is consistent with previous reports linking the oxidative degradation of nitrophenolic compounds to hydroxyl radical attack and subsequent electron transfer processes [205].

Importantly, no cathodic peak was detected during the reverse scan from +0.70 to +0.40 V, suggesting that the electrochemical transformation pathway of pNP is irreversible under the applied conditions.

The appearance of both electroactive intermediates (such as hydroquinone) and non-electroactive products has been widely documented during photoelectrocatalytic (PEC) degradation of phenolic pollutants, further supporting the proposed degradation mechanism of pNP under UV-irradiated TiO$_2$ [205,206]. Overall, the gradual increase in anodic current with irradiation time, together with the stable anodic peak at +0.55 V, confirms that TiO$_2$

photoactivation drives the oxidative degradation of pNP via •OH-mediated pathways [205,206].

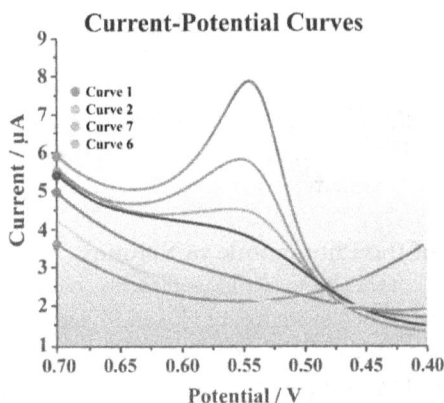

Figure 5-9: HDPV profiles of 1×10^{-4} mol·L^{-1} pNP in 0.1 mol·L^{-1} H$_2$SO$_4$ at a platinum-ring electrode under varying UV illumination times (0–60 min); rotation rate: 1600 rpm [175].

3-3 Photoelectrocatalytic Degradation Using Pure and Nitrogen-Doped TiO$_2$ Nanotube Arrays

Photoelectrocatalysis (PEC) has emerged as a simple yet highly effective approach to mitigate the rapid recombination of photogenerated electron–hole pairs, a key limitation in conventional photocatalytic systems. In particular, applying PEC principles to TiO$_2$ nanotube arrays (NTAs) has been shown to significantly enhance their degradation performance.

As shown in Figure 5-10, applying a low bias potential to TiO$_2$ nanotube arrays (NTAs) enhances the separation and transport of photogenerated charge carriers, thereby reducing recombination. Upon ultraviolet (UV) illumination, electrons are excited from the valence band (VB) to the conduction band (CB) of TiO$_2$ and subsequently flow through an external circuit toward the counter electrode. Concurrently, the photogenerated holes remain on the surface of the TiO$_2$ NTAs, where they can participate in oxidation reactions.

These separated charge carriers participate in redox reactions that lead to the generation of reactive species. Through redox reactions at the electrode interfaces, electrons in the conduction band (CB) reduce molecular oxygen to form superoxide radicals (O$_2$•$^-$), while holes in the valence band (VB) oxidize water molecules, generating hydroxyl radicals (•OH). These highly reactive species play a pivotal role in the oxidative degradation of organic pollutants [207].

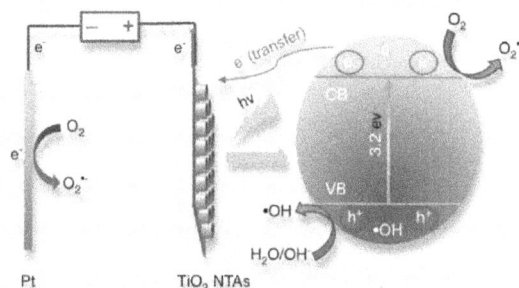

Figure 5-10: Schematic Diagram for Photoelectrocatalytic Degradation of Pollutants under UV Irradiation [208].

The application of an external bias potential in photoelectrochemical (PEC) systems plays a critical role in minimizing the recombination of photogenerated electron–hole pairs. Compared to their unmodified counterparts, TiO_2 nanotube arrays (NTAs) modified through surface or structural alterations exhibit significantly enhanced degradation of organic contaminants, particularly under UV and visible light irradiation. Among various modifications, nitrogen doping is especially effective, as it reduces the band gap of TiO_2, promotes more efficient charge carrier separation, and improves visible-light absorption. These modifications significantly improve photochemical conversion efficiency and PEC performance relative to pristine TiO_2 NTAs [209].

Antonius B. Aritonang et al. synthesized highly ordered nitrogen-doped TiO_2 nanotube arrays (N-TNTAs) via a one-step anodization process conducted at 40 V for 1 hour in an electrolyte containing ammonium fluoride (NH_4F), water, and triethylene glycol. The resulting samples were annealed in a nitrogen atmosphere at 450 °C for 3 hours. The photoelectrochemical (PEC) performance of both N-TNTA and undoped TNTA electrodes was assessed through the degradation of methylene blue (MB) in aqueous solution. Experiments were carried out in 0.1 M KOH under vigorous stirring, using an initial MB concentration of 10 mg/L in a 30 mL solution at pH 5.5. Prior to illumination, electrodes were immersed in the MB solution in the dark for 30 minutes to establish adsorption–desorption equilibrium. During the PEC process, 2 mL aliquots were collected every 10 minutes, analyzed using a Shimadzu 2450 UV–Vis spectrophotometer at 655 nm, and then returned to the reactor. Under visible-light irradiation, the N-TNTA electrode achieved an MB degradation efficiency of 89%, markedly surpassing the performance of the undoped TNTA photoanode. The higher reaction rate constant observed for the nitrogen-doped electrode further confirmed its enhanced photoelectrocatalytic activity [86].

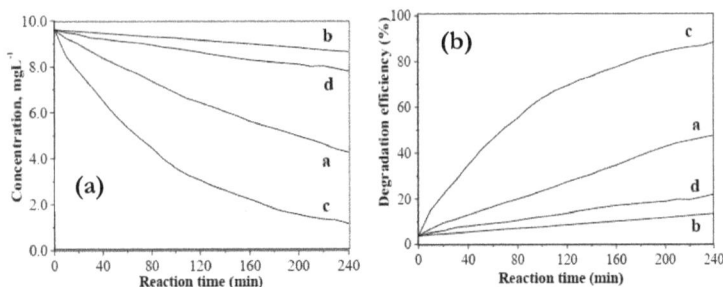

Figure 5-11: (A) MB concentration and (B) degradation efficiency under various conditions; (C) PEC of N-TNTA and (D) PEC of TNTA under visible light at 0.2 V bias [86].

4- Photoelectrochemical cells

A photoelectrochemical (PEC) cell is a specific type of electrochemical cell that generates electric current and voltage through photon absorption at one or both electrodes. These systems typically utilize semiconductor materials as photoactive electrodes. As illustrated in Figure 5-12, incident light on the semiconductor electrode generates electron–hole pairs. Charge separation at the semiconductor–electrolyte interface is driven by band bending and energy level alignment, which promote the directional movement of electrons toward the electrolyte and holes toward the back contact or reactive surface [5].

One of the key advantages of PEC cells over traditional photocatalytic systems lies in the spatial separation of oxidation and reduction reactions. In photocatalytic systems, both photogenerated electrons and holes participate in reactions at the same location, which increases the likelihood of recombination. In contrast, PEC systems allow oxidation and reduction to occur at separate electrodes, thereby improving charge separation and overall process efficiency [5].

The performance of PEC cells is significantly influenced by the morphology of the semiconductor electrode. Nanoparticle films typically suffer from short carrier lifetimes and low diffusion coefficients due to their disordered crystalline structure and abundant grain boundaries. However, in well-aligned nanotube arrays, parallel channels within the structure enable better electrolyte penetration and facilitate oxidant transport. Furthermore, the highly ordered wall structure of TiO_2 nanotubes provides direct pathways for electron transfer. By tuning nanotube dimensions, the internal void space can be optimized, which improves charge carrier lifetime and enhances PEC activity. Their mechanical durability and strong adhesion to the substrate make nanotube electrodes a superior choice for PEC configurations [106].

One of the most promising applications of photoelectrochemical (PEC) cells is hydrogen production through solar-driven water splitting. As a clean and renewable energy carrier, hydrogen offers a sustainable alternative to fossil fuels. The pioneering study by Fujishima and Honda in 1972 was the first to demonstrate the feasibility of photoinduced water splitting using a TiO_2 photoanode under light irradiation [108]. For this process to occur efficiently, the semiconductor's band edge positions must straddle the redox potentials of water, thereby enabling both hydrogen and oxygen evolution. As shown in Figure 5-12b, photogenerated electrons drive water reduction and hydrogen evolution at the photocathode, while holes at the photoanode participate in the oxygen evolution reaction [5].

In addition to energy generation, photoelectrochemical (PEC) cells are widely utilized for the degradation of dyes and various organic pollutants. Upon illumination, oxidation reactions occur at the photoelectrode surface, leading to the breakdown of harmful compounds [5,210]. Introducing an external bias significantly enhances photoelectrocatalytic efficiency by promoting charge carrier separation. Compared to nanoparticulate films, TiO_2 nanotube-based electrodes exhibit superior charge transport and improved light-harvesting capabilities. This trend is illustrated in Figure 5-13, which highlights the superior degradation efficiency achieved through PEC compared to conventional photocatalytic or electrochemical methods [210].

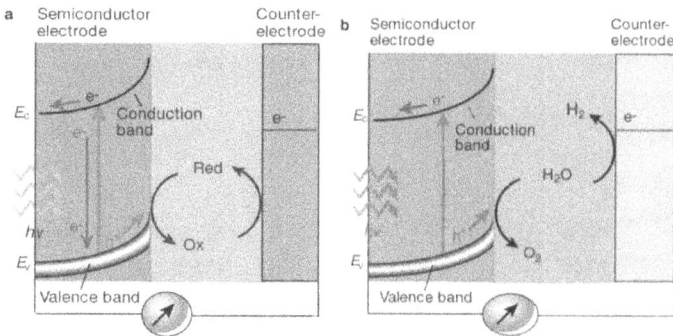

Figure 5-12: Schematic of a photoelectrochemical cell a) to carry out oxidation and reduction photoelectrocatalytic reactions and b) to produce hydrogen [108].

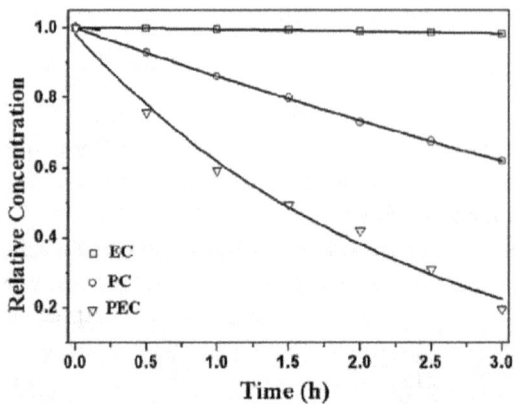

Figure 5-13: Comparison of the decomposition rate of phenolic compounds by electrochemical (EC), photocatalytic (PC) and electrophotocatalytic (PEC) methods [111].

CHAPTER 6: ENSITY FUNCTIONAL THEORY— FROM MANY-BODY QUANTUM MECHANICS TO PRACTICAL ELECTRONIC-STRUCTURE MODELING

1- Introduction

Understanding the electronic structure of materials—from isolated molecules to extended solids—requires solving the many-electron Schrödinger equation, a task that becomes intractable as particle number increases. Density Functional Theory (DFT) provides a transformative framework: instead of working with the exponentially complex many-electron wavefunction, it reformulates the problem in terms of the three-dimensional electron density [211,212]. The Hohenberg–Kohn theorems established that (i) the ground-state density uniquely determines all properties of an interacting electron system and (ii) the correct density minimizes a universal energy functional [211]. Building on these principles, Kohn and Sham introduced an auxiliary system of non-interacting electrons moving in an effective potential that reproduces the same ground-state density as the real system [212]. This approach made DFT both conceptually rigorous and computationally feasible.

In practice, the accuracy of DFT relies on approximations to the exchange–correlation (XC) functional. The Local Density Approximation (LDA), based on the homogeneous electron gas, offers a first-level description, while Generalized Gradient Approximations (GGAs) incorporate density gradients for improved accuracy in inhomogeneous systems [213–215]. More advanced approaches—including hybrid functionals, meta-GGA formulations, and DFT+U corrections—extend DFT's applicability to strongly correlated materials, localized d/f electrons, and van der Waals interactions [216,217]. Spin-polarized extensions further enable the study of magnetism by treating spin-resolved densities as fundamental variables [217]. Despite limitations such as band-gap underestimation and challenges with strongly correlated systems, DFT remains the most widely used tool in computational materials science, chemistry, and condensed matter physics due to its balance of accuracy and efficiency [213,214,216–218].

For oxide semiconductors like TiO_2, DFT has been particularly influential. It provides critical insights into how polymorphs (anatase vs. rutile), defects (oxygen vacancies, Ti^{3+} centers), and doping (e.g., N, C, S, F) modify band edges, introduce mid-gap states, and impact carrier dynamics [4,109,219]. These theoretical insights directly complement experimental findings in photocatalysis, solar energy conversion, and electrochemical storage, where nitrogen-doped TiO_2 nanotubes exemplify the synergy between nanoscale architecture and electronic-structure engineering [109,219]. As such, DFT not only bridges quantum theory with real materials but also guides the rational design of next-generation photocatalysts, electrodes, and energy devices.

2- Density Functional Theory

Understanding the behavior of complex physical systems—particularly those involving many interacting particles—often requires quantum mechanical approaches. While exact solutions are feasible for simple systems, such as isolated atoms or diatomic molecules, solving systems with more particles becomes mathematically intractable. As a result, modern theoretical and computational methods like Density Functional Theory (DFT) have been developed to address the electronic structure of such many-body systems efficiently. These methods often rely on specific approximations and are selected based on the nature of the problem being addressed. One key concept relevant to many systems—excluding dynamic structures such as molecules and crystals—is the operational velocity, which plays a fundamental role in system behavior. This concept is frequently utilized alongside the Cohen–Shem method to characterize subsystems within larger, multi-component models [220].

2-1 Many-particle systems

In quantum mechanics, many-particle systems consist of atoms, molecules, clusters, or solids in which numerous particles interact simultaneously. These systems exhibit complex behavior that cannot be described using simple analytical models. To analyze them, the first step is to construct the system's Hamiltonian—a mathematical operator representing the total energy of the system. The Hamiltonian includes both kinetic and potential energy terms and dictates how the system evolves over time:

$$\hat{H} = \sum_i \frac{-h^2}{2m_e} \nabla_i^2 + \sum_I \frac{-h^2}{2m_I} \nabla_I^2 \qquad (6\text{-}1)$$
$$- \sum_{i,I} \frac{-z_I e^2}{|\vec{r_i} - \vec{R_I}|} + \frac{1}{2} \sum_{i \neq j} \frac{e^2}{|\vec{r_i} - \vec{r_j}|}$$
$$+ \frac{1}{2} \sum_{i \neq j} \frac{z_I z_J e^2}{|\vec{R_I} - \vec{R_J}|}$$

In this model, electrons—each with mass m and charge e—are located at positions r_i, while ions, which are much heavier, have masses M_i and charges $Z_i e$, and are positioned at R_i. To simplify the formulation of the Hamiltonian, relativistic effects are ignored, and the influence of any external electric or magnetic fields is neglected. The various terms in the Hamiltonian correspond, respectively, to the kinetic energy of electrons, the kinetic energy of ions, electron–ion interactions, electron–electron interactions, and ion–ion interactions [220,221].

2-2 Born-Oppenheimer approximation

In many-particle systems, the Hamiltonian includes both electronic and nuclear contributions. However, because nuclei are about 2000 times heavier than electrons, their motion is significantly slower. This difference allows physicists to simplify the problem by applying the Born–Oppenheimer approximation.

According to this approximation:
a) The nuclei are treated as stationary while solving the electronic Schrödinger equation.
b) The electrons are assumed to remain in their ground state as the nuclear positions slowly change.

This simplification enables the separation of electronic and nuclear motions, allowing the complex Hamiltonian to be reduced to an electron-only form. As a result, the positions of the nuclei are treated as fixed parameters when analyzing electronic behavior.

Despite lacking a rigorous mathematical proof, the Born–Oppenheimer approximation remains a foundational concept in quantum simulations, offering a practical means to decouple electronic and nuclear motions in many-body systems [222] Applying this approximation reduces equation (6-2) to the following simplified Hamiltonian:

$$\hat{H} = \sum_i \frac{-h^2}{2m_e} \nabla_i^2 - \sum_{i,I} \frac{-z_I e^2}{\left|\vec{r}_i - \vec{R}_I\right|} + \frac{1}{2} \sum_{i \neq j} \frac{e^2}{\left|\vec{r}_i - \vec{r}_j\right|} \qquad (6\text{-}2)$$

When applying the Born–Oppenheimer approximation, the total particle Hamiltonian is simplified into an effective multi-electron Hamiltonian. However, even with this simplification, solving the electronic component of multi-particle systems remains a complex and demanding task. The primary difficulty arises from electron–electron interactions, which introduce many-body effects that cannot be neglected.

In the absence of electron–electron interactions, the system could be treated as a collection of independent single-electron Hamiltonians, significantly simplifying the calculations. However, the Coulomb repulsion between electrons prevents such decoupling, making the system inherently correlated. Despite their computational complexity, electron–electron interactions play a fundamental role in explaining a broad spectrum of physical phenomena, including the metal–insulator transition, the Kondo effect, heavy fermion behavior, high-temperature superconductivity, and the quantum Hall effect.

For instance, the t–J model, which incorporates electron correlation and quantum fluctuations, is widely used to investigate the metal–nonconductor transition. This model demonstrates how the interplay between electronic interactions and quantum dynamics governs the electronic phases of matter.

It is also important to note that within the Born–Oppenheimer approximation, electron–phonon interactions are inherently excluded, as the motion of electrons is decoupled from that of the ions. This separation implies that phonon-related effects—such as vibrational contributions to conductivity or lattice interactions—are not captured in purely electron-based models [220,222].

2-3 Independent electron approximation

To simplify the challenge of modeling electron–electron interactions in multi-electron systems, physicists often use the independent electron approximation. In this model, electrons are not treated as interacting with one another directly. Instead, each electron moves in an average electrostatic field created by all other electrons—this is known as the Hartree approximation.

The resulting Hartree potential replaces complex pairwise interactions with a single, effective mean-field. This approach significantly reduces computational complexity while still capturing the dominant effects of electron repulsion. Although it simplifies the system, the Hartree model does not fully account for all quantum mechanical effects, especially electron correlation and exchange—areas where further refinement is needed.

$$\hat{H} = \sum_i \frac{-h^2}{2m_e} \nabla_i^2 + \sum_i^V ext(\vec{r}_i) \qquad (6\text{-}3a)$$

$$V_H(\vec{r}) = \int d^3\vec{r}\, \frac{e\, n(?)}{\vec{r} - \vec{r}} \qquad (6\text{-}3b)$$

In this formulation, the electron density is represented by *n(r)* in the potential term *U*, while the electron–ion interaction energy appears as the second term of the Hamiltonian. In this context, ions are treated as external fields influencing the electronic system.

This Hamiltonian can be decomposed into a sum of single-electron components, leading to what is known as the independent-electron approximation. Under this approximation, each electron is considered to move independently within the effective Hartree potential generated by the averaged presence of all other electrons.

Although the explicit electron–electron interaction is excluded, the Hartree Hamiltonian cannot fully reproduce the true many-electron wave function. Nevertheless, because a significant portion of electron–electron interactions is effectively captured by the Hartree potential, the resulting electron density and total energy often closely approximate those derived from the exact many-body solution. A more refined and comprehensive Hamiltonian based on the independent-electron framework was subsequently developed to address the limitations of this basic approach [220].

2-4 Hohenberg-Cohen issues

Density Functional Theory (DFT) is based on two fundamental theorems introduced by Hohenberg and Kohn.

The first theorem states that the ground-state electron density uniquely determines all properties of a quantum system, including the external potential and total Hamiltonian. In other words, if you know the exact electron density, you can—at least in principle—reconstruct all the physical characteristics of the system without needing its full wavefunction.

This is powerful because wavefunctions for N-electron systems exist in a 3N-dimensional space, making direct computation extremely difficult. In contrast, electron density depends only on three spatial variables, offering a far simpler route for modeling complex systems.

The second theorem establishes that the total energy of a system is a functional of the electron density, and that the correct ground-state density is the one that minimizes this energy functional. This is the basis of the variational principle in DFT: the system's true ground-state energy is obtained by finding the density that yields the lowest possible energy [222]. This fundamental relationship is formally expressed as follows [220].

$$E[n(\vec{r})] = T[n(\vec{r})] + \int d^3\vec{r}n(\vec{r})V_{ext}(\vec{r}) + E_{int}[n(\vec{r})] \qquad (4)$$

The terms comprising the total energy functional correspond, respectively, to the kinetic energy of the electrons, the electron–ion interaction, and the electron–electron interaction. According to the second fundamental theorem of Density Functional Theory (DFT), once the total energy functional is established, the ground-state electron density serves as the central quantity from which all physical properties of the system can, in principle, be derived. However, a significant challenge arises from the fact that the exact forms of the kinetic energy and electron–electron interaction functionals are not explicitly known. This uncertainty limits the direct application of DFT and necessitates the use of approximations to model these terms accurately. The following section discusses various strategies that have been developed to address this issue, focusing on the construction of practical exchange–correlation functionals that approximate these unknown contributions.

2-5 Cohen-Sham method

In 1965, Kohn and Sham revolutionized Density Functional Theory by proposing a practical solution to the many-body problem. Rather than attempting to solve the full interacting-electron equations, they introduced an imaginary system of non-interacting electrons that would yield the same ground-state electron density as the real system.

The core idea is that, although these electrons do not interact with each other, their behavior is governed by a specially designed effective potential. Solving the Schrödinger equation for this simpler system can accurately reproduce the electron density of the original, interacting system.

This effective potential—called the Kohn–Sham potential—has three components:

1) The external potential (usually from atomic nuclei),
2) The classical Hartree term (electron–electron repulsion),
3) The exchange–correlation potential, which accounts for the quantum mechanical interactions beyond classical effects.

With this setup, the many-body problem is reduced to a set of single-electron equations, each describing one electron's motion under the combined influence of the effective potential. While the electrons are mathematically treated as non-interacting, the functional form of the exchange–correlation term ensures that many-body effects are still captured accurately.

$$E_{ks} = T_{ks} + E_{ext} + E_{eff} \qquad (6\text{-}5)$$

In this framework, T_{KS} represents the kinetic energy of non-interacting electrons within the Kohn–Sham auxiliary system. The term E_{ext} captures the interaction between these electrons and the external potential—typically arising from the nuclei in the system.

Meanwhile, E_{eff} accounts for the effects of electron–electron interactions, but in an indirect way. Instead of explicitly including every interaction, these effects are modeled using an effective potential. This term is often broken down into two essential components: the Hartree energy, which describes the classical electrostatic repulsion between electrons, and the exchange–correlation energy, which captures the more subtle, quantum mechanical corrections that go beyond the mean-field approach.

By organizing the total energy in this way, the Kohn–Sham method strikes a smart balance—it includes the essential physics of many-body interactions while keeping the problem computationally manageable. This balance is one of the key reasons why DFT has become so powerful and widely used in modern quantum chemistry and materials science [220].

$$T_{ks} = \frac{-h^2}{2m} \sum_{i,\sigma} <\varphi_i^\sigma |\nabla^2| \varphi_i^\sigma> = \frac{-h^2}{2m} \sum_{i,\sigma} |\nabla \varphi_i^\sigma|^2 \qquad (6\text{-}6a)$$

$$E_{ext} = \int d^3r \, V_{ext}(r) n(r) \qquad n(r) = \sum_{i,\sigma} |\varphi_i^\sigma|^2 \qquad (6\text{-}6b)$$

According to Hartree's theory, which provides an approximation for the potential created by electron–electron interactions, the Hartree energy is considered to make up the main portion of E_{eff} in the Kohn–Sham framework. In simpler terms, this means that the classical electrostatic repulsion between electrons is the dominant contributor to the effective interaction energy.

However, the picture isn't complete without one more piece. The remaining effects—those subtle quantum mechanical behaviors that go beyond the classical model—are captured in a correction term known as the exchange–correlation energy.

In summary, the total energy of the Kohn–Sham system—as expressed in Equation (6-5)—consists of four main components: the kinetic energy of the non-interacting electrons, their interaction with the external potential, the Hartree energy, and the exchange–correlation energy, which encapsulates the more subtle aspects of electron correlation.

$$E_{ks} = \frac{-h^2}{2m} \sum_{i,\sigma} |\nabla \varphi_i^\sigma|^2 \tag{6-7}$$

$$+ \int d^3\vec{r} V_{ext}(\vec{r}) n(\vec{r})$$

$$+ \frac{1}{2} e^2 \int d^3\vec{r} d^3\vec{r} \frac{n(\vec{r})n(\vec{r})}{|\vec{r} - \vec{r}|} + E_{xc}[n(\vec{r})]$$

The exchange–correlation energy originates from the difference between the total energy of the real interacting electron system—as defined by the functional in Equation (6-4)—and that of the non-interacting auxiliary system in the Kohn–Sham framework.

$$T_{ks} + E_{ext} + E_H + E_{xc} = T + E_{ext} + E_{int} \rightarrow E_{xc} = (T\text{-}T_{ks}) + (E_{int} - E_H) \tag{6-8}$$

The exchange–correlation energy (E_{xc}) comprises two principal contributions: the difference between the true electron–electron interaction energy and the classical Hartree energy, and the difference between the kinetic energy of the interacting (real) system and that of the non-interacting Kohn–Sham system. Although E_{xc} represents a vital component of the total energy in Density Functional Theory (DFT), its exact form remains unknown. Consequently, several approximation methods—such as the Local Density Approximation (LDA) and the Generalized Gradient Approximation (GGA)—have been developed to estimate it with reasonable accuracy in practical calculations [222].

According to the Rayleigh–Ritz variational principle in quantum mechanics [222], the Kohn–Sham Hamiltonian is derived from the Kohn–Sham energy functional and is formulated under the constraint of a fixed number of particles, with the additional requirement that the wave functions remain orthonormal. This principle guarantees that the ground-state energy, obtained by minimizing the energy functional with respect to trial wave functions, serves as an upper bound to the exact ground-state energy.

$$\frac{\delta}{\delta \varphi_i^{\sigma^*}} \left(E_{ks} - \varepsilon \int n(r) d^3 r \right) = 0 \qquad (6\text{-}9)$$

Based on equation (6-7), the individual components of the Kohn–Sham energy functional can be calculated as follows:

$$\frac{\delta T_{ks}}{\delta \varphi_i^{\sigma^*}} = \frac{\delta}{\delta \varphi_i^{\sigma^*}} \left(\frac{-h^2}{2m} \sum_{k.a} < \mu_k^a | \nabla^2 | \varphi_k^a \right) = \frac{-h^2}{2m} \nabla^2 \varphi_i^{\sigma} \qquad (6\text{-}10)$$

Since the remaining terms in the energy functional are expressed as functionals of the electron density, the chain rule of functional differentiation is employed to derive their corresponding functional derivatives.

$$\frac{\delta}{\delta \varphi_i^{\sigma^*}} = \frac{\delta n}{\delta \varphi_i^{\sigma^*}} \frac{\delta}{\delta n} = \frac{\delta}{\delta \varphi_i^{\sigma^*}} \left(\sum_{k.a} \varphi_k^{a^*} \varphi_k^a \right) \frac{\delta}{\delta n} = \varphi_i^{\sigma} \frac{\delta}{\delta n} \qquad (6\text{-}11)$$

Based on Equation (6-9), the functional expressions for the Hartree energy and the electron–external potential interaction can be derived as follows:

$$\frac{\delta}{\delta \varphi_i^{\sigma^*}} (E_{ext} + E_H) = \varphi_i^{\sigma} \frac{\delta}{\delta n} (E_{ext} + E_H) \qquad (6\text{-}12)$$

$$= \varphi_i^{\sigma} (V_{ext} \, e^2 \int \frac{n(\vec{r}) d^3 \vec{r}}{|\vec{r} - \vec{r}|} = \varphi_i^{\sigma} (V_{ext}(\vec{r}) + V_H(\vec{r}))$$

Accordingly, the exchange–correlation energy is defined by the following expression:

$$\frac{\delta}{\delta \varphi_i^{\sigma^*}} (E_{xc}[n(r)] = \varphi_i^{\sigma}(r) \frac{\delta E_{xc}}{\delta n} = \varphi_i^{\sigma}(r) V_{xc}(r) \qquad (6\text{-}13)$$

It is evident that the exchange–correlation potential, defined as $(V_{xc} = \frac{\delta E_{xc}}{\delta n})$, is obtained by taking the functional derivative of the exchange–correlation energy with respect to the electron density. This result is derived by substituting the relevant expressions into equation (6-9) of the Kohn–Sham single-particle equation set, and by incorporating the outcomes of equations (6-10), (6-12), and (6-13).

$$\left(\frac{-h^2}{2m} \nabla^2 + V_{ext}(r) + V_H(r) + V_{xc}(r) \right) \varphi_i^{\sigma}(r) = \varepsilon \varphi_i^{\sigma}(r) \qquad (6\text{-}14)$$

Finally, the Kohn–Sham Hamiltonian is formally defined by the following expression:

$$H = \frac{-h^2}{2m}\nabla^2 + V_{ext}(r) + V_H(r) + V_{xc}(r)$$

(6-15)

The exchange-correlation energy (E_{xc}) can, in principle, be accurately determined by first evaluating the exchange–correlation potential (V_{xc}), and subsequently solving the single-particle Kohn–Sham equations. This procedure allows for an exact description of the initial state of the many-body system, including its ground-state energy and electron density. However, the exchange-correlation energy remains the most challenging component of the total energy functional, as it cannot be derived exactly and must instead be approximated. To this end, several approximation schemes—such as the Local Density Approximation (LDA) and the Generalized Gradient Approximation (GGA) [223] —have been developed to estimate this term with reasonable accuracy.

The simplest of these methods, the LDA, is derived from an in-depth analysis of a homogeneous electron gas within the framework of Hartree-Fock theory. It assumes that the exchange–correlation energy at each point in space depends exclusively on the local value of the electron density.

$$E_{xc} \approx \int d^3 r n(r)\, \varepsilon_x^{homo}(n(r))$$

(6-16)

In the expression (U)n above, the electron density nnn appears explicitly, and the integration is performed over the entire crystal. The exchange–correlation energy per electron in a homogeneous electron gas is denoted by ε_{xc}^{homo}. When applied locally, this leads to the following relation, presented as equation (6-16).

$$E_{xc}^{GGA} = \int d^3 r n(r) \left[\varepsilon_x^{homo}(n(r)) + \varepsilon_c^{homo}(n(r))\right]$$

(6-17)

The function $\varepsilon_x^{homo}(n(r))$, which represents the exchange energy per electron in a homogeneous electron gas as a function of the local electron density $n(r)$, is also approximated using the following expression:

$$\varepsilon_x^{homo}(n(r)) = -\frac{3}{4}(\frac{6}{4}n(r))^{-\frac{1}{3}}$$

(6-18)

The analysis of the correlation energy $\varepsilon_c^{homo}(n(r))$ is somewhat more complex than that of the exchange energy $\varepsilon_x^{homo}(n(r))$. Its accurate value is obtained through the quantum Monte Carlo method. This approach allows for precise energy calculations, which are then fitted into other governing expressions. The resulting parameterizations are used in various exchange-correlation functionals, with several prominent models described in references [213,221].

$$\varepsilon_c^{pz} = \begin{cases} -0.1423/(1 + 1.9539\sqrt{r_s} + 0.33r_s) & r_s \geq 1 \\ -0.0480 + 0.0311 L_n r_n - 0.0116 r_s + 0.0020 r_s L_n r_n & r_s < 1 \end{cases} \qquad (6\text{-}19)$$

The Generalized Gradient Approximation (GGA) extends the Local Density Approximation (LDA) by incorporating electron density gradients, thereby improving accuracy in systems with non-uniform electron distributions. Numerous relationships and functional forms have been developed based on this approach. In GGA, the exchange-correlation energy at each point in space is determined not only by the local electron density but also by its spatial variation (i.e., its gradient). As a result, the exchange-correlation energy takes the following general form:

$$E_{xc}^{GGA} = \int d^3 r n(r) \varepsilon_{xc}(\text{n.}|\nabla n|) = \int d^3 r n(r) \varepsilon_x^{homo} F_{xc}(\text{n.}|\nabla n|) \qquad (6\text{-}20)$$

that the function F_{xc} is dimensionless, and we shall have:

$$F_{xc}(n.|\nabla n|) = F_x(n.|\nabla n|) + F_c(n.|\nabla n|) \qquad (6\text{-}21)$$

The F_c calculations are more intricate than the F_x calculations. Despite having a negligible impact in the equation above, the correlation energy contribution pales in comparison to the exchange energy. Nonetheless, several functions are available to determine the Fc value. However, because F_x is more important than F_c when Exc is offered, just the names of functions connected to F_x are mentioned, such as pw91 and pw96 [223].

The Cohen-Shem Hamiltonian cannot be employed to verify the properties of magnetic systems because it does not take spin into account Instead, a wide range of materials and compounds' various properties are examined through their use. In practice, modified relations of density functional theory—which is akin to assigning it to systems including an external magnetic field—are employed in place of correlation exchange complex functions. In this context, the properties of the system are described in terms of the spin-resolved components of the ground-state electron density, rather than the total ground-state density. When applying Density Functional Theory (DFT) within a spin-polarized framework, it is important to note that only the exchange–correlation functional is generalized to depend on the spin components of the electron density. In contrast, both the Hartree energy and the external potential remain functionals of the total electron density [221,223].

The calculation of the correlation energy F_c is more complex than that of the exchange energy F_x. Although its impact in the above equation is relatively small, the contribution of correlation energy is generally minor compared to the exchange energy. Nonetheless, several functions have been developed to estimate F_c. However, due to the dominant role of F_x in the total exchange-correlation energy E_{xc}, only functions associated with F_x, such as PW91 and PW96 [223], are commonly referenced.

List of Figures

List of Tables

Refrence

[1] "Lütjering, G. and Williams, J.C. (2007) Titanium. 2nd Edition, Springer, Berlin Heidelberg. - References - Scientific Research Publishing." Retrieved 20 September 2025. https://www.scirp.org/reference/referencespapers?referenceid=2144 844

[2] Donachie, M. J. ., "Titanium : A Technical Guide," 2000, p. 381. Retrieved 20 September 2025

[3] Zhang, H., and Banfield, J. F., "Understanding Polymorphic Phase Transformation Behavior during Growth of Nanocrystalline Aggregates: Insights from TiO2," *Journal of Physical Chemistry B*, Vol. 104, No. 15, 2000, pp. 3481–3487. https://doi.org/10.1021/JP000499J

[4] Diebold, U., "The Surface Science of Titanium Dioxide," *Surface Science Reports*, Vol. 48, Nos. 5–8, 2003, pp. 53–229. https://doi.org/10.1016/S0167-5729(02)00100-0

[5] Grätzel, M., "Photoelectrochemical Cells," *Nature*, Vol. 414, No. 6861, 2001, pp. 338–344. https://doi.org/10.1038/35104607

[6] Hu, M. Z., Lai, P., Bhuiyan, M. S., Tsouris, C., Gu, B., Parans Paranthaman, M., Gabitto, J., and Harrison, L., "Synthesis and Characterization of Anodized Titanium-Oxide Nanotube Arrays," *Journal of Materials Science*, Vol. 44, No. 11, 2009, pp. 2820–2827. https://doi.org/10.1007/S10853-009-3372-4/METRICS

[7] Wang, X., Li, Z., Shi, J., and Yu, Y., "One-Dimensional Titanium Dioxide Nanomaterials: Nanowires, Nanorods, and Nanobelts," *Chemical Reviews*, Vol. 114, No. 19, 2014, pp. 9346–9384. https://doi.org/10.1021/CR400633S/ASSET/CR400633S.FP.PNG_V03

[8] Regonini, D., Bowen, C. R., Jaroenworaluck, A., and Stevens, R., "A Review of Growth Mechanism, Structure and Crystallinity of Anodized TiO2 Nanotubes," *Materials Science & Engineering R - Reports*, Vol. 74, No. 12, 2013, pp. 377–406. https://doi.org/10.1016/J.MSER.2013.10.001

[9] Kulkarni, M., Mazare, A., and Schmuki, P., "Influence of Anodization Parameters on Morphology of TiO 2 Nanostructured Surfaces,"

Advanced Materials Letters Adv. Mater. Lett, Vol. 7, No. 1, 2016, pp. 23–28. https://doi.org/10.5185/amlett.2016.6156

[10]   Regonini, D., Groff, A., Sorarù, G. D., and Clemens, F. J., "Photoelectrochemical Study of Anodized TiO2 Nanotubes Prepared Using Low and High H2O Contents," *Electrochimica Acta*, Vol. 186, 2015,   pp.   101–111. https://doi.org/10.1016/J.ELECTACTA.2015.10.162

[11]   Zhang, Q., Ma, L., Xu, X., Shao, M., Huang, J., Han, L., and Li, W., "The Influence of Oxidation Time on the Morphologies of TiO2 Nanostructures," *Journal of Nanoscience and Nanotechnology*, Vol. 14, No. 4, 2014, pp. 3262–3265. https://doi.org/10.1166/JNN.2014.8557

[12]   Beranek, R., Hildebrand, H., and Schmuki, P., "Self-Organized Porous Titanium Oxide Prepared in H2SO4/HF Electrolytes," *Electrochemical and Solid-State Letters*, Vol. 6, No. 3, 2003, p. B12. https://doi.org/10.1149/1.1545192/XML

[13]   Apolinário, A., Sousa, C. T., Ventura, J., Costa, J. D., Leitão, D. C., Moreira, J. M., Sousa, J. B., Andrade, L., Mendes, A. M., and Araújo, J. P., "The Role of the Ti Surface Roughness in the Self-Ordering of TiO2 Nanotubes: A Detailed Study of the Growth Mechanism," *Journal of Materials Chemistry A*, Vol. 2, No. 24, 2014, pp. 9067–9078. https://doi.org/10.1039/C4TA00871E

[14]   Indira, K., Mudali, U. K., Nishimura, T., and Rajendran, N., "A Review on TiO2 Nanotubes: Influence of Anodization Parameters, Formation Mechanism, Properties, Corrosion Behavior, and Biomedical Applications," *Journal of Bio- and Tribo-Corrosion 2015 1:4*, Vol. 1, No. 4, 2015, pp. 1–22. https://doi.org/10.1007/S40735-015-0024-X

[15]   Guo, C., Hu, F., Li, C. M., and Shen, P. K., "Direct Electrochemistry of Hemoglobin on Carbonized Titania Nanotubes and Its Application in a Sensitive Reagentless Hydrogen Peroxide Biosensor," *Biosensors and Bioelectronics*, Vol. 24, No. 4, 2008, pp. 819–824. https://doi.org/10.1016/J.BIOS.2008.07.007

[16]   Yang, Y., Wang, X., and Li, L., "Synthesis and Growth Mechanism of Graded TiO2 Nanotube Arrays by Two-Step Anodization," *Materials Science and Engineering: B*, Vol. 149, No. 1, 2008, pp. 58–62.

https://doi.org/10.1016/J.MSEB.2007.12.006

[17]   Indira, K., Ningshen, S., Mudali, U. K., and Rajendran, N., "Effect of Anodization Parameters on the Structural Morphology of Titanium in Fluoride Containing Electrolytes," *Materials Characterization*, Vol. 71, 2012, pp. 58–65. https://doi.org/10.1016/J.MATCHAR.2012.06.005

[18]   Tsuchiya, H., MacAk, J. M., Ghicov, A., Taveira, L., and Schmuki, P., "Self-Organized Porous TiO2 and ZrO2 Produced by Anodization," *Corrosion Science*, Vol. 47, No. 12, 2005, pp. 3324–3335. https://doi.org/10.1016/J.CORSCI.2005.05.041

[19]   Macák, J. M., Tsuchiya, H., and Schmuki, P., "High-Aspect-Ratio TiO2 Nanotubes by Anodization of Titanium," *Angewandte Chemie International Edition*, Vol. 44, No. 14, 2005, pp. 2100–2102. https://doi.org/10.1002/ANIE.200462459

[20]   Macak, J. M., Hildebrand, H., Marten-Jahns, U., and Schmuki, P., "Mechanistic Aspects and Growth of Large Diameter Self-Organized TiO2 Nanotubes," *Journal of Electroanalytical Chemistry*, Vol. 621, No. 2, 2008,   pp.   254–266. https://doi.org/10.1016/J.JELECHEM.2008.01.005

[21]   Sun, K. C., Chen, Y. C., Kuo, M. Y., Wang, H. W., Lu, Y. F., Chung, J. C., Liu, Y. C., and Zeng, Y. Z., "Synthesis and Characterization of Highly Ordered TiO2 Nanotube Arrays for Hydrogen Generation via Water Splitting," *Materials Chemistry and Physics*, Vol. 129, Nos. 1–2, 2011, pp. 35–39. https://doi.org/10.1016/J.MATCHEMPHYS.2011.03.081

[22]   Sreekantan, S., Saharudin, K. A., and Wei, L. C., "Formation of TiO2 Nanotubes via Anodization and Potential Applications for Photocatalysts, Biomedical Materials, and Photoelectrochemical Cell," *IOP Conference Series: Materials Science and Engineering*, Vol. 21, No. 1, 2011, p. 012002. https://doi.org/10.1088/1757-899X/21/1/012002

[23]   Seong, W. M., Kim, D. H., Park, I. J., Park, G. Do, Kang, K., Lee, S., and Hong, K. S., "Roughness of Ti Substrates for Control of the Preferred Orientation of TiO2 Nanotube Arrays as a New Orientation Factor," *Journal of Physical Chemistry C*, Vol. 119, No. 23, 2015, pp. 13297–13305. https://doi.org/10.1021/ACS.JPCC.5B02371/SUPPL_FILE/JP5B02

371_SI_001.PDF

[24] "(PDF) Photocatalytic TiO2 Thin Films for Air Cleaning: Effect of Facet Orientation, Chemical Functionalization, and Reaction Conditions." Retrieved 21 June 2025. https://www.researchgate.net/publication/291137056_Photocatalytic_TiO2_thin_films_for_air_cleaning_Effect_of_facet_orientation_chemical_functionalization_and_reaction_conditions

[25] Tizazu, G., El-Zubir, O., Brueck, S. R. J., Lidzey, D. G., Leggett, G. J., and Lopez, G. P., "Large Area Nanopatterning of Alkylphosphonate Self-Assembled Monolayers on Titanium Oxide Surfaces by Interferometric Lithography," *Nanoscale*, Vol. 3, No. 6, 2011, pp. 2511–2516. https://doi.org/10.1039/C0NR00994F

[26] Pishkar, N., Ghoranneviss, M., Ghorannevis, Z., and Akbari, H., "Study of the Highly Ordered TiO2 Nanotubes Physical Properties Prepared with Two-Step Anodization," *Results in Physics*, Vol. 9, 2018, pp. 1246–1249. https://doi.org/10.1016/J.RINP.2018.02.009

[27] Schuisky, M., Hårsta, A., Aidla, A., Kukli, K., Kiisler, A.-A., and Aarik, J., "Atomic Layer Chemical Vapor Deposition of TiO[Sub 2] Low Temperature Epitaxy of Rutile and Anatase," *Journal of The Electrochemical Society*, Vol. 147, No. 9, 2000, p. 3319. https://doi.org/10.1149/1.1393901/XML

[28] Kominami, H., Murakami, S. Y., Kato, J. I., Kera, Y., and Ohtani, B., "Correlation between Some Physical Properties of Titanium Dioxide Particles and Their Photocatalytic Activity for Some Probe Reactions in Aqueous Systems," *Journal of Physical Chemistry B*, Vol. 106, No. 40, 2002, pp. 10501–10507. https://doi.org/10.1021/JP0147224;ISSUE:ISSUE:10.1021/JPCBFK.2002.106.ISSUE-40;WGROUP:STRING:ACHS

[29] Ohtani, B., Zhang, S., Handa, J., Kajiwara, H., Nishimoto, S., and Kagiya, T., "Photocatalytic Activity of Titanium(IV) Oxide Prepared from Titanium(IV) Tetra-2-Propoxide: Reaction in Aqueous Silver Salt Solutions," *Journal of Photochemistry and Photobiology A: Chemistry*, Vol. 64, No. 2, 1992, pp. 223–230. https://doi.org/10.1016/1010-6030(92)85109-8

[30] Chen, X., and Mao, S. S., "Titanium Dioxide Nanomaterials: Synthesis, Properties, Modifications and Applications," *Chemical Reviews*, Vol. 107, No. 7, 2007, pp. 2891–2959. https://doi.org/10.1021/CR0500535/ASSET/CR0500535.FP.PNG_V03

[31] Roy, P., Berger, S., and Schmuki, P., "TiO2 Nanotubes: Synthesis and Applications," *Angewandte Chemie International Edition*, Vol. 50, No. 13, 2011, pp. 2904–2939. https://doi.org/10.1002/ANIE.201001374

[32] Mor, G. K., Varghese, O. K., Paulose, M., Shankar, K., and Grimes, C. A., "A Review on Highly Ordered, Vertically Oriented TiO2 Nanotube Arrays: Fabrication, Material Properties, and Solar Energy Applications," *Solar Energy Materials and Solar Cells*, Vol. 90, No. 14, 2006, pp. 2011–2075. https://doi.org/10.1016/J.SOLMAT.2006.04.007

[33] Shankar, K., Mor, G. K., Prakasam, H. E., Yoriya, S., Paulose, M., Varghese, O. K., and Grimes, C. A., "Highly-Ordered TiO2 Nanotube Arrays up to 220 Mm in Length: Use in Water Photoelectrolysis and Dye-Sensitized Solar Cells," *Nanotechnology*, Vol. 18, No. 6, 2007. https://doi.org/10.1088/0957-4484/18/6/065707

[34] "Kinetics and Mechanism of the Anatase/Rutile Transformation, as Catalyzed by Ferric Oxide and Reducing Conditions | American Mineralogist | GeoScienceWorld." Retrieved 7 May 2025. https://pubs.geoscienceworld.org/msa/ammin/article-abstract/57/1-2/10/542569/Kinetics-and-mechanism-of-the-anatase-rutile

[35] Kominami, H., Kohno, M., and Kera, Y., "Synthesis of Brookite-Type Titanium Oxide Nano-Crystals in Organic Media," *Journal of Materials Chemistry*, Vol. 10, No. 5, 2000, pp. 1151–1156. https://doi.org/10.1039/A908528I

[36] Wehrenfennig, C., Palumbiny, C. M., Snaith, H. J., Johnston, M. B., Schmidt-Mende, L., and Herz, L. M., "Fast Charge-Carrier Trapping in TiO2 Nanotubes," *Journal of Physical Chemistry C*, Vol. 119, No. 17, 2015, pp. 9159–9168. https://doi.org/10.1021/ACS.JPCC.5B01827

[37] Chen, Z. H., Tang, Y. B., Liu, C. P., Leung, Y. H., Yuan, G. D., Chen, L. M., Wang, Y. Q., Bello, I., Zapien, J. A., Zhang, W. J., Lee, C. S., and Lee, S. T., "Vertically Aligned ZnO Nanorod Arrays Sentisized with

Gold Nanoparticles for Schottky Barrier Photovoltaic Cells," *Journal of Physical Chemistry C*, Vol. 113, No. 30, 2009, pp. 13433–13437. https://doi.org/10.1021/JP903153W;PAGEGROUP:STRING:PUBL ICATION

[38]   Feng, X., Shankar, K., Varghese, O. K., Paulose, M., Latempa, T. J., and Grimes, C. A., "Vertically Aligned Single Crystal TiO 2 Nanowire Arrays Grown Directly on Transparent Conducting Oxide Coated Glass: Synthesis Details and Applications," *Nano Letters*, Vol. 8, No. 11, 2008,   pp.   3781–3786. https://doi.org/10.1021/NL802096A/SUPPL_FILE/NL802096A_S I_001.PDF

[39]   Dubey, M., He, H., Dubey, M., and He, H., "Morphological and Photovoltaic Studies of TiO2 NTs for High Efficiency Solar Cells," *Scanning Electron Microscopy*, 2012. https://doi.org/10.5772/36332

[40]   Han, L., Koide, N., Chiba, Y., Islam, A., Komiya, R., Fuke, N., Fukui, A., and Yamanaka, R., "Improvement of Efficiency of Dye-Sensitized Solar Cells by Reduction of Internal Resistance," *Applied Physics Letters*, Vol.   86,   No.   21,   2005,   pp.   1–3. https://doi.org/10.1063/1.1925773/931474

[41]   Lee, K. M., Suryanarayanan, V., and Ho, K. C., "The Influence of Surface Morphology of TiO2 Coating on the Performance of Dye-Sensitized Solar Cells," *Solar Energy Materials and Solar Cells*, Vol. 90, No. 15,   2006,   pp.   2398–2404. https://doi.org/10.1016/J.SOLMAT.2006.03.034

[42]   Kang, S. H., Kim, H. S., Kim, J. Y., and Sung, Y. E., "An Investigation on Electron Behavior Employing Vertically-AlignedTiO2 Nanotube Electrodes for Dye-Sensitized Solar Cells," *Nanotechnology*, Vol. 20, No. 35,   2009,   p.   355307.   https://doi.org/10.1088/0957-4484/20/35/355307

[43]   Peighambardoust, N. S., and Nasirpouri, F., "Manipulating Morphology, Pore Geometry and Ordering Degree of TiO2 Nanotube Arrays by Anodic Oxidation," *Surface and Coatings Technology*, Vol. 235, 2013,   pp.   727–734. https://doi.org/10.1016/J.SURFCOAT.2013.08.058

[44] Mor, G. K., Shankar, K., Paulose, M., Varghese, O. K., and Grimes, C. A., "Enhanced Photocleavage of Water Using Titania Nanotube Arrays," *Nano Letters*, Vol. 5, No. 1, 2005, pp. 191–195. https://doi.org/10.1021/NL048301K/ASSET/IMAGES/MEDIUM/NL048301KN00001.GIF

[45] Allam, N. K., and Grimes, C. A., "Effect of Cathode Material on the Morphology and Photoelectrochemical Properties of Vertically Oriented TiO2 Nanotube Arrays," *Solar Energy Materials and Solar Cells*, Vol. 92, No. 11, 2008, pp. 1468–1475. https://doi.org/10.1016/J.SOLMAT.2008.06.007

[46] Asahi, R., Morikawa, T., Ohwaki, T., Aoki, K., and Taga, Y., "Visible-Light Photocatalysis in Nitrogen-Doped Titanium Oxides," *Science*, Vol. 293, No. 5528, 2001, pp. 269–271. https://doi.org/10.1126/SCIENCE.1061051;CSUBTYPE:STRING:SPECIAL;PAGE:STRING:ARTICLE/CHAPTER

[47] Mollavali, M., Falamaki, C., and Rohani, S., "Preparation of Multiple-Doped TiO2 Nanotube Arrays with Nitrogen, Carbon and Nickel with Enhanced Visible Light Photoelectrochemical Activity via Single-Step Anodization," *International Journal of Hydrogen Energy*, Vol. 40, No. 36, 2015, pp. 12239–12252. https://doi.org/10.1016/J.IJHYDENE.2015.07.069

[48] Khan, S. U. M., Al-Shahry, M., and Ingler, W. B., "Efficient Photochemical Water Splitting by a Chemically Modified N-TiO2," *Science*, Vol. 297, No. 5590, 2002, pp. 2243–2245. https://doi.org/10.1126/SCIENCE.1075035,

[49] Chen, X., Liu, L., Yu, P. Y., and Mao, S. S., "Increasing Solar Absorption for Photocatalysis with Black Hydrogenated Titanium Dioxide Nanocrystals," *Science*, Vol. 331, No. 6018, 2011, pp. 746–750. https://doi.org/10.1126/SCIENCE.1200448/SUPPL_FILE/CHEN.X.SOM.PDF

[50] Zhang, H., Lv, X., Li, Y., Wang, Y., and Li, J., "P25-Graphene Composite as a High Performance Photocatalyst," *ACS Nano*, Vol. 4, No. 1, 2009, pp. 380–386. https://doi.org/10.1021/NN901221K

[51] Salari, M., Konstantinov, K., and Liu, H. K., "Enhancement of the

Capacitance in TiO2 Nanotubes through Controlled Introduction of Oxygen Vacancies," *Journal of Materials Chemistry*, Vol. 21, No. 13, 2011, pp. 5128–5133. https://doi.org/10.1039/C0JM04085A

[52] Zhou, H., and Zhang, Y., "Electrochemically Self-Doped TiO2 Nanotube Arrays for Supercapacitors," *Journal of Physical Chemistry C*, Vol. 118, No. 11, 2014, pp. 5626–5636. https://doi.org/10.1021/JP4082883

[53] Wu, L., Li, F., Xu, Y., Zhang, J. W., Zhang, D., Li, G., and Li, H., "Plasmon-Induced Photoelectrocatalytic Activity of Au Nanoparticles Enhanced TiO2 Nanotube Arrays Electrodes for Environmental Remediation," *Applied Catalysis B: Environmental*, Vol. 164, 2015, pp. 217–224. https://doi.org/10.1016/J.APCATB.2014.09.029

[54] Raj, C. C., Sundheep, R., and Prasanth, R., "Enhancement of Electrochemical Capacitance by Tailoring the Geometry of TiO2 Nanotube Electrodes," *Electrochimica Acta*, Vol. 176, 2015, pp. 1214–1220. https://doi.org/10.1016/J.ELECTACTA.2015.07.052

[55] Raj, C. C., and Prasanth, R., " Review—Advent of TiO 2 Nanotubes as Supercapacitor Electrode ," *Journal of The Electrochemical Society*, Vol. 165, No. 9, 2018, pp. E345–E358. https://doi.org/10.1149/2.0561809JES

[56] Yuan, R., Zhou, B., Hua, D., and Shi, C., "Enhanced Photocatalytic Degradation of Humic Acids Using Al and Fe Co-Doped TiO2 Nanotubes under UV/Ozonation for Drinking Water Purification," *Journal of Hazardous Materials*, Vol. 262, 2013, pp. 527–538. https://doi.org/10.1016/J.JHAZMAT.2013.09.012

[57] Ren, F., Li, H., Wang, Y., and Yang, J., "Enhanced Photocatalytic Oxidation of Propylene over V-Doped TiO2 Photocatalyst: Reaction Mechanism between V5+ and Single-Electron-Trapped Oxygen Vacancy," *Applied Catalysis B: Environmental*, Vols. 176–177, 2015, pp. 160–172. https://doi.org/10.1016/J.APCATB.2015.03.050

[58] Szkoda, M., Siuzdak, K., and Lisowska-Oleksiak, A., "Non-Metal Doped TiO2 Nanotube Arrays for High Efficiency Photocatalytic Decomposition of Organic Species in Water," *Physica E: Low-dimensional Systems and Nanostructures*, Vol. 84, 2016, pp. 141–145. https://doi.org/10.1016/J.PHYSE.2016.06.004

[59]   Mohamed, A. E. R., Barghi, S., and Rohani, S., "N- and C-Modified TiO2 Nanotube Arrays: Enhanced Photoelectrochemical Properties and Effect of Nanotubes Length on Photoconversion Efficiency," *Nanomaterials 2018, Vol. 8, Page 198*, Vol. 8, No. 4, 2018, p. 198. https://doi.org/10.3390/NANO8040198

[60]   Wilke, K., and Breuer, H. D., "The Influence of Transition Metal Doping on the Physical and Photocatalytic Properties of Titania," *Journal of Photochemistry and Photobiology A: Chemistry*, Vol. 121, No. 1, 1999, pp. 49–53. https://doi.org/10.1016/S1010-6030(98)00452-3

[61]   Sakthivel, S., and Kisch, H., "Photocatalytic and Photoelectrochemical Properties of Nitrogen-Doped Titanium Dioxide," *ChemPhysChem*, Vol. 4,   No.   5,   2003,   pp.   487–490. https://doi.org/10.1002/CPHC.200200554;CTYPE:STRING:JOUR NAL

[62]   Park, J. H., Kim, S., and Bard, A. J., "Novel Carbon-Doped TiO 2 Nanotube Arrays with High Aspect Ratios for Efficient Solar Water Splitting," *Nano Letters*, Vol. 6, No. 1, 2006, pp. 24–28. https://doi.org/10.1021/NL051807Y/ASSET/IMAGES/MEDIUM /NL051807YN00001.GIF

[63]   Umebayashi, T., Yamaki, T., Itoh, H., and Asai, K., "Band Gap Narrowing of Titanium Dioxide by Sulfur Doping," *Applied Physics Letters*, Vol. 81, No. 3, 2002, pp. 454–456. https://doi.org/10.1063/1.1493647

[64]   Yang, C., Wang, Z., Lin, T., Yin, H., Lü, X., Wan, D., Xu, T., Zheng, C., Lin, J., Huang, F., Xie, X., and Jiang, M., "Core-Shell Nanostructured 'Black' Rutile Titania as Excellent Catalyst for Hydrogen Production Enhanced by Sulfur Doping," *Journal of the American Chemical Society*, Vol. 135, No. 47, 2013, pp. 17831–17838. https://doi.org/10.1021/JA4076748/SUPPL_FILE/JA4076748_SI_0 01.PDF

[65]   Lin, T., Yang, C., Wang, Z., Yin, H., Lü, X., Huang, F., Lin, J., Xie, X., and Jiang, M., "Effective Nonmetal Incorporation in Black Titania with Enhanced Solar Energy Utilization," *Energy & Environmental Science*, Vol. 7, No. 3, 2014, pp. 967–972. https://doi.org/10.1039/C3EE42708K

[66]   Geim, A. K., and Novoselov, K. S., "The Rise of Graphene," *Nature Materials 2007 6:3*, Vol. 6, No. 3, 2007, pp. 183–191. https://doi.org/10.1038/nmat1849

[67]   Wang, C., Meng, D., Sun, J., Memon, J., Huang, Y., and Geng, J., "Graphene Wrapped TiO2 Based Catalysts with Enhanced Photocatalytic Activity," *Advanced Materials Interfaces*, Vol. 1, No. 4, 2014, p. 1300150. https://doi.org/10.1002/ADMI.201300150

[68]   Dong, J., Han, J., Liu, Y., Nakajima, A., Matsushita, S., Wei, S., and Gao, W., "Defective Black TiO2 Synthesized via Anodization for Visible-Light Photocatalysis," *ACS Applied Materials and Interfaces*, Vol. 6, No. 3, 2014, pp. 1385–1388. https://doi.org/10.1021/AM405549P/SUPPL_FILE/AM405549P_S I_001.PDF

[69]   Kim, C., Kim, S., Lee, J., Kim, J., and Yoon, J., "Capacitive and Oxidant Generating Properties of Black-Colored TiO2 Nanotube Array Fabricated by Electrochemical Self-Doping," *ACS Applied Materials and Interfaces*, Vol. 7, No. 14, 2015, pp. 7486–7491. https://doi.org/10.1021/ACSAMI.5B00123

[70]   Yang, Y., and Hoffmann, M. R., "Synthesis and Stabilization of Blue-Black TiO2 Nanotube Arrays for Electrochemical Oxidant Generation and Wastewater Treatment," *Environmental Science and Technology*, Vol. 50, No. 21, 2016, pp. 11888–11894. https://doi.org/10.1021/ACS.EST.6B03540/SUPPL_FILE/ES6B03 540_SI_001.PDF

[71]   Bergamonti, L., Predieri, G., Paz, Y., Fornasini, L., Lottici, P. P., and Bondioli, F., "Enhanced Self-Cleaning Properties of N-Doped TiO2 Coating for Cultural Heritage," *Microchemical Journal*, Vol. 133, 2017, pp. 1–12. https://doi.org/10.1016/J.MICROC.2017.03.003

[72]   Mittal, A., Mari, B., Sharma, S., Kumari, V., Maken, S., Kumari, K., and Kumar, N., "Non-Metal Modified TiO 2 : A Step towards Visible Light Photocatalysis," *Journal of Materials Science: Materials in Electronics*, Vol. 30, No. 4, 2019, pp. 3186–3207. https://doi.org/10.1007/S10854-018-00651-9

[73]   Preethi, L. K., Antony, R. P., Mathews, T., Walczak, L., and Gopinath,

C. S., "A Study on Doped Heterojunctions in TiO2 Nanotubes: An Efficient Photocatalyst for Solar Water Splitting," *Scientific Reports 2017 7:1*, Vol. 7, No. 1, 2017, pp. 1–15. https://doi.org/10.1038/s41598-017-14463-0

[74] Cheng, Z. L., and Han, S., "Preparation and Photoelectrocatalytic Performance of N-Doped TiO2/NaY Zeolite Membrane Composite Electrode Material," *Water science and technology : a journal of the International Association on Water Pollution Research*, Vol. 73, No. 3, 2016, pp. 486–492. https://doi.org/10.2166/WST.2015.505

[75] May Ix, L. A., Estrella González, A., Cipagauta-Díaz, S., and Gómez, R., "Effective Electron–Hole Separation over N-Doped TiO2 Materials for Improved Photocatalytic Reduction of 4-Nitrophenol Using Visible Light," *Journal of Chemical Technology & Biotechnology*, Vol. 95, No. 10, 2020, pp. 2694–2706. https://doi.org/10.1002/JCTB.6365

[76] Siuzdak, K., Szkoda, M., Sawczak, M., and Lisowska-Oleksiak, A., "Novel Nitrogen Precursors for Electrochemically Driven Doping of Titania Nanotubes Exhibiting Enhanced Photoactivity," *New Journal of Chemistry*, Vol. 39, No. 4, 2015, pp. 2741–2751. https://doi.org/10.1039/C5NJ00127G

[77] Huang, L., Fu, W., Fu, X., Zong, B., Liu, H., Bala, H., Wang, X., Sun, G., Cao, J., and Zhang, Z., "Facile and Large-Scale Preparation of N Doped TiO2 Photocatalyst with High Visible Light Photocatalytic Activity," *Materials Letters*, Vol. 209, 2017, pp. 585–588. https://doi.org/10.1016/J.MATLET.2017.08.092

[78] Jiang, X., Wang, Y., and Pan, C., "High Concentration Substitutional N-Doped TiO2 Film: Preparation, Characterization, and Photocatalytic Property," *Journal of the American Ceramic Society*, Vol. 94, No. 11, 2011, pp. 4078–4083. https://doi.org/10.1111/J.1551-2916.2011.04692.X

[79] Mohan, L., Anandan, C., and Rajendran, N., "Effect of Plasma Nitriding on Structure and Biocompatibility of Self-Organised TiO2 Nanotubes on Ti–6Al–7Nb," *RSC Advances*, Vol. 5, No. 52, 2015, pp. 41763–41771. https://doi.org/10.1039/C5RA05818J

[80] Hou, X., Liu, F., Yao, K., Ma, H., Deng, J., Li, D., and Liao, B., "Photoelectrical Properties of Nitrogen Doped TiO2 Nanotubes by

Anodic Oxidation of N+ Implanted Ti Foils," *Materials Letters*, Vol. 124, 2014, pp. 101–104. https://doi.org/10.1016/J.MATLET.2014.03.050

[81]   Ghicov, A., Macak, J. M., Tsuchiya, H., Kunze, J., Haeublein, V., Frey, L., and Schmuki, P., "Ion Implantation and Annealing for an Efficient N-Doping of TiO2 Nanotubes," *Nano Letters*, Vol. 6, No. 5, 2006, pp. 1080–1082. https://doi.org/10.1021/NL0600979

[82]   Macak, J. M., Ghicov, A., Hahn, R., Tsuchiya, H., and Schmuki, P., "Photoelectrochemical Properties of N-Doped Self-Organized Titania Nanotube Layers with Different Thickness," *Journal of Materials Research*, Vol. 21, No. 11, 2006, pp. 2824–2828. https://doi.org/10.1557/JMR.2006.0344

[83]   Zhang, X., Zhou, J., Gu, Y., and Fan, D., "Visible-Light Photocatalytic Activity of N-Doped TiO2 Nanotube Arrays on Acephate Degradation," *Journal of Nanomaterials*, Vol. 2015, No. 1, 2015, p. 527070. https://doi.org/10.1155/2015/527070

[84]   Wang, Y., Zhu, L., Ba, N., Gao, F., and Xie, H., "Effects of NH4F Quantity on N-Doping Level, Photodegradation and Photocatalytic H2 Production Activities of N-Doped TiO2 Nanotube Array Films," *Materials Research Bulletin*, Vol. 86, 2017, pp. 268–276. https://doi.org/10.1016/J.MATERRESBULL.2016.10.031

[85]   Li, S., Lin, S., Liao, J., Pan, N., Li, D., and Li, J., "Nitrogen-Doped TiO2 Nanotube Arrays with Enhanced Photoelectrochemical Property," *International Journal of Photoenergy*, Vol. 2012, 2012. https://doi.org/10.1155/2012/794207

[86]   Du, K., Liu, G., Chen, X., -, al, Sun, Q., Hong, Y., Zang, T., Rajashekhar, H., Vrushabendrakumar, D., Masud Rana, M., Aritonang, A. B., Surahman, H., Krisnandi, Y. K., and Gunlazuardi, J., "Photo-Electro-Catalytic Performance of Highly Ordered Nitrogen Doped TiO2 Nanotubes Array Photoanode," *IOP Conference Series: Materials Science and Engineering*, Vol. 172, No. 1, 2017, p. 012005. https://doi.org/10.1088/1757-899X/172/1/012005

[87]   Antony, R. P., Mathews, T., Ajikumar, P. K., Krishna, D. N., Dash, S., and Tyagi, A. K., "Electrochemically Synthesized Visible Light Absorbing Vertically Aligned N-Doped TiO2 Nanotube Array Films,"

Materials Research Bulletin, Vol. 47, No. 12, 2012, pp. 4491–4497. https://doi.org/10.1016/J.MATERRESBULL.2012.09.061

[88] Divyasri, Y. V., Lakshmana Reddy, N., Lee, K., Sakar, M., Navakoteswara Rao, V., Venkatramu, V., Shankar, M. V., and Gangi Reddy, N. C., "Optimization of N Doping in TiO2 Nanotubes for the Enhanced Solar Light Mediated Photocatalytic H2 Production and Dye Degradation," *Environmental Pollution*, Vol. 269, 2021, p. 116170. https://doi.org/10.1016/J.ENVPOL.2020.116170

[89] Sun, L., Cai, J., Wu, Q., Huang, P., Su, Y., and Lin, C., "N-Doped TiO2 Nanotube Array Photoelectrode for Visible-Light-Induced Photoelectrochemical and Photoelectrocatalytic Activities," *Electrochimica Acta*, Vol. 108, 2013, pp. 525–531. https://doi.org/10.1016/J.ELECTACTA.2013.06.149

[90] Yuan, B., Wang, Y., Bian, H., Shen, T., Wu, Y., and Chen, Z., "Nitrogen Doped TiO2 Nanotube Arrays with High Photoelectrochemical Activity for Photocatalytic Applications," *Applied Surface Science*, Vol. 280, 2013, pp. 523–529. https://doi.org/10.1016/J.APSUSC.2013.05.021

[91] Subramaniam, V. P., Tang, Y. X., and Gong, D. G., "Nitrogen-Doped TiO2 Nanotube Array Films with Enhanced Photocatalytic Activity under Various Light Sources Nitrogen-Doped TiO2 Nanotube Array Films with Enhanced Photocatalytic Activity under Various Light Sources Nitrogen-Doped TiO 2 Nanotube Array Films with Enhanced Photocatalytic Activity under Various Light Sources," *Journal of Hazardous Materials*, Vol. 184, No. 3, 2010, pp. 855–863. https://doi.org/10.1016/j.jhazmat.2010.08.121

[92] Mahy, J. G., Cerfontaine, V., Poelman, D., Devred, F., Gaigneaux, E. M., Heinrichs, B., and Lambert, S. D., "Highly Efficient Low-Temperature N-Doped TiO2 Catalysts for Visible Light Photocatalytic Applications," *Materials 2018, Vol. 11, Page 584*, Vol. 11, No. 4, 2018, p. 584. https://doi.org/10.3390/MA11040584

[93] Di Valentin, C., Finazzi, E., Pacchioni, G., Selloni, A., Livraghi, S., Paganini, M. C., and Giamello, E., "N-Doped TiO2: Theory and Experiment," *Chemical Physics*, Vol. 339, Nos. 1–3, 2007, pp. 44–56.

https://doi.org/10.1016/J.CHEMPHYS.2007.07.020

[94] Asahi, R., Morikawa, T., Irie, H., and Ohwaki, T., "Nitrogen-Doped Titanium Dioxide as Visible-Light-Sensitive Photocatalyst: Designs, Developments, and Prospects," *Chemical Reviews*, Vol. 114, No. 19, 2014, pp. 9824–9852. https://doi.org/10.1021/CR5000738/ASSET/CR5000738.FP.PNG_V03

[95] Dong, F., Zhao, W., Wu, Z., and Guo, S., "Band Structure and Visible Light Photocatalytic Activity of Multi-Type Nitrogen Doped TiO2 Nanoparticles Prepared by Thermal Decomposition," *Journal of Hazardous Materials*, Vol. 162, Nos. 2–3, 2009, pp. 763–770. https://doi.org/10.1016/J.JHAZMAT.2008.05.099

[96] Hou, X., Aitola, K., Jiang, H., Lund, P. D., and Li, Y., "Reduced TiO2 Nanotube Array as an Excellent Cathode for Hydrogen Evolution Reaction in Alkaline Solution," *Catalysis Today*, Vol. 402, 2022, pp. 3–9. https://doi.org/10.1016/J.CATTOD.2021.12.009

[97] Fang, D., Luo, Z., Huang, K., and Lagoudas, D. C., "Effect of Heat Treatment on Morphology, Crystalline Structure and Photocatalysis Properties of TiO2 Nanotubes on Ti Substrate and Freestanding Membrane," *Applied Surface Science*, Vol. 257, No. 15, 2011, pp. 6451–6461. https://doi.org/10.1016/J.APSUSC.2011.02.037

[98] Lin, J., Liu, X., Zhu, S., Liu, Y., and Chen, X., "Anatase TiO2 Nanotube Powder Film with High Crystallinity for Enhanced Photocatalytic Performance," *Nanoscale Research Letters*, Vol. 10, No. 1, 2015, p. 110. https://doi.org/10.1186/S11671-015-0814-6

[99] Dikova, T., Hashim, D. P., and Mintcheva, N., "Morphology and Structure of TiO2 Nanotube/Carbon Nanostructure Coatings on Titanium Surfaces for Potential Biomedical Application," *Materials 2024, Vol. 17, Page 1290*, Vol. 17, No. 6, 2024, p. 1290. https://doi.org/10.3390/MA17061290

[100] Le, P. H., Hieu, L. T., Lam, T. N., Hang, N. T. N., Van Truong, N., Tuyen, L. T. C., Phong, P. T., and Leu, J., "Enhanced Photocatalytic Performance of Nitrogen-Doped TiO2 Nanotube Arrays Using a Simple Annealing Process," *Micromachines 2018, Vol. 9, Page 618*, Vol. 9,

No. 12, 2018, p. 618. https://doi.org/10.3390/MI9120618

[101] Pisarek, M., Krawczyk, M., Gurgul, M., Zaraska, L., Bieńkowski, K., Hołdyński, M., and Solarska, R., "Plasma-Assisted N-Doped TiO2 Nanotube Array as an Active UV-Vis Photoanode," *ACS Applied Nano Materials*, Vol. 6, No. 12, 2023, pp. 10351–10364. https://doi.org/10.1021/ACSANM.3C01278/ASSET/IMAGES/LARGE/AN3C01278_0010.JPEG

[102] Pisarek, M., Krawczyk, M., Hołdyński, M., and Lisowski, W., "Plasma Nitriding of TiO2 Nanotubes: N-Doping in Situ Investigations Using XPS," *ACS Omega*, Vol. 5, No. 15, 2020, pp. 8647–8658. https://doi.org/10.1021/ACSOMEGA.0C00094/ASSET/IMAGES/LARGE/AO0C00094_0002.JPEG

[103] Natarajan, T. S., Mozhiarasi, V., and Tayade, R. J., "Nitrogen Doped Titanium Dioxide (N-TiO2): Synopsis of Synthesis Methodologies, Doping Mechanisms, Property Evaluation and Visible Light Photocatalytic Applications," *Photochem 2021, Vol. 1, Pages 371-410*, Vol. 1, No. 3, 2021, pp. 371–410. https://doi.org/10.3390/PHOTOCHEM1030024

[104] Zhang, Z., Cui, Z., Xu, Y., Ghazzal, M. N., Colbeau-Justin, C., Pan, D., and Wu, W., "A Facile Strategy for the Preparation of N-Doped TiO2 with Oxygen Vacancy via the Annealing Treatment with Urea," *Nanomaterials*, Vol. 14, No. 10, 2024, p. 818. https://doi.org/10.3390/NANO14100818/S1

[105] Zhou, X., Häublein, V., Liu, N., Nguyen, N. T., Zolnhofer, E. M., Tsuchiya, H., Killian, M. S., Meyer, K., Frey, L., and Schmuki, P., "TiO2 Nanotubes: N-Ion Implantation at Low-Dose Provides Noble-Metal-Free Photocatalytic H2-Evolution Activity," *Angewandte Chemie - International Edition*, Vol. 55, No. 11, 2016, pp. 3763–3767. https://doi.org/10.1002/anie.201511580

[106] Mor, G. K., Varghese, O. K., Paulose, M., Shankar, K., and Grimes, C. A., "A Review on Highly Ordered, Vertically Oriented TiO2 Nanotube Arrays: Fabrication, Material Properties, and Solar Energy Applications," *Solar Energy Materials and Solar Cells*, Vol. 90, No. 14, 2006, pp. 2011–2075. https://doi.org/10.1016/J.SOLMAT.2006.04.007

[107] Appadurai, T., Subramaniyam, C. M., Kuppusamy, R., Karazhanov, S., and Subramanian, B., "Electrochemical Performance of Nitrogen-Doped TiO2 Nanotubes as Electrode Material for Supercapacitor and Li-Ion Battery," *Molecules 2019, Vol. 24, Page 2952*, Vol. 24, No. 16, 2019, p. 2952. https://doi.org/10.3390/MOLECULES24162952

[108] Fujishima, A., and Honda, K., "Electrochemical Photolysis of Water at a Semiconductor Electrode," *Nature 1972 238:5358*, Vol. 238, No. 5358, 1972, pp. 37–38. https://doi.org/10.1038/238037a0

[109] Chen, X., Liu, L., and Huang, F., "Black Titanium Dioxide (TiO2) Nanomaterials," *Chemical Society Reviews*, Vol. 44, No. 7, 2015, pp. 1861–1885. https://doi.org/10.1039/C4CS00330F

[110] Grätzel, M., "Dye-Sensitized Solar Cells," *Journal of Photochemistry and Photobiology C: Photochemistry Reviews*, Vol. 4, No. 2, 2003, pp. 145–153. https://doi.org/10.1016/S1389-5567(03)00026-1

[111] Naldoni, A., Allieta, M., Santangelo, S., Marelli, M., Fabbri, F., Cappelli, S., Bianchi, C. L., Psaro, R., and Dal Santo, V., "Effect of Nature and Location of Defects on Bandgap Narrowing in Black TiO2 Nanoparticles," *Journal of the American Chemical Society*, Vol. 134, No. 18, 2012, pp. 7600–7603. https://doi.org/10.1021/JA3012676

[112] Mor, G. K., Shankar, K., Paulose, M., Varghese, O. K., and Grimes, C. A., "Enhanced Photocleavage of Water Using Titania Nanotube Arrays," *Nano Letters*, Vol. 5, No. 1, 2005, pp. 191–195. https://doi.org/10.1021/NL048301K/ASSET/IMAGES/MEDIUM/NL048301KN00001.GIF

[113] Roy, P., Berger, S., and Schmuki, P., "TiO2 Nanotubes: Synthesis and Applications," *Angewandte Chemie International Edition*, Vol. 50, No. 13, 2011, pp. 2904–2939. https://doi.org/10.1002/ANIE.201001374

[114] Mazierski, P., Nischk, M., Gołkowska, M., Lisowski, W., Gazda, M., Winiarski, M. J., Klimczuk, T., and Zaleska-Medynska, A., "Photocatalytic Activity of Nitrogen Doped TiO2 Nanotubes Prepared by Anodic Oxidation: The Effect of Applied Voltage, Anodization Time and Amount of Nitrogen Dopant," *Applied Catalysis B: Environmental*, Vol. 196, 2016, pp. 77–88. https://doi.org/10.1016/J.APCATB.2016.05.006

[115] Zhou, Y., Liu, Y., Liu, P., Zhang, W., Xing, M., and Zhang, J., "A Facile Approach to Further Improve the Substitution of Nitrogen into Reduced TiO2−x with an Enhanced Photocatalytic Activity," *Applied Catalysis B: Environmental*, Vols. 170–171, 2015, pp. 66–73. https://doi.org/10.1016/J.APCATB.2015.01.036

[116] Kumar Jena, B., and Retna Raj, C., "Morphology Dependent Electrocatalytic Activity of Au Nanoparticles," *Electrochemistry Communications*, Vol. 10, No. 6, 2008, pp. 951–954. https://doi.org/10.1016/J.ELECOM.2008.04.023

[117] Du, K., Liu, G., Chen, X., -, al, Sun, Q., Hong, Y., Zang, T., Elysabeth, T., and Sri Redjeki, A., "Synthesis of N Doped Titania Nanotube Arrays Photoanode Using Urea as Nitrogen Precursor for Photoelectrocatalytic Application," *IOP Conference Series: Materials Science and Engineering*, Vol. 509, No. 1, 2019, p. 012144. https://doi.org/10.1088/1757-899X/509/1/012144

[118] Lee, J. H., Mun, S. J., Lee, S. Y., and Park, S. J., "Promoted Charge Separation and Specific Surface Area via Interlacing of N-Doped Titanium Dioxide Nanotubes on Carbon Nitride Nanosheets for Photocatalytic Degradation of Rhodamine B," *Nanotechnology Reviews*, Vol. 11, No. 1, 2022, pp. 1592–1605. https://doi.org/10.1515/NTREV-2022-0085/ASSET/GRAPHIC/J_NTREV-2022-0085_FIG_008.JPG

[119] Ghorbani, F., and Asl, S. K., "Modifying TiO2 Nanotube Using N-Doping and Electrochemical Reductive Doping as a Supercapacitor Electrode," *Journal of Ultrafine Grained and Nanostructured Materials*, Vol. 54, No. 2, 2021, pp. 131–140. https://doi.org/10.22059/JUFGNSM.2021.02.02

[120] Yang, B., She, Y., Zhang, C., Kang, S., Zhou, J., and Hu, W., "Nitrogen Doped Intercalation TiO2/TiN/Ti3C2Tx Nanocomposite Electrodes with Enhanced Pseudocapacitance," *Nanomaterials*, Vol. 10, No. 2, 2020, p. 345. https://doi.org/10.3390/NANO10020345

[121] Akbar, H., Ali, A., Mohammad, S., Anjum, F., Ahmad, A., Afzal, A. M., Albaqami, M. D., Mohammad, S., and Choi, J. R., "Exploring the Potential of Nitrogen-Doped Graphene in ZnSe-TiO2 Composite

Materials for Supercapacitor Electrode," *Molecules 2024, Vol. 29, Page 2103*, Vol. 29, No. 9, 2024, p. 2103. https://doi.org/10.3390/MOLECULES29092103

[122] Sun, L., Chen, Y., Zhuo, K., Sun, D., Liu, J., and Wang, J., "Nitrogen-Doped Ti3C2Tx/TiO2-Graphene Hybrid Materials with 3D Interconnected Structure Compatible with Ionic Liquid-Based Electrolyte for High Performance Supercapacitors," *Electrochimica Acta*, Vol. 466, 2023. https://doi.org/10.1016/J.ELECTACTA.2023.143005

[123] Bakar, S. A., and Ribeiro, C., "Nitrogen-Doped Titanium Dioxide: An Overview of Material Design and Dimensionality Effect over Modern Applications," *Journal of Photochemistry and Photobiology C: Photochemistry Reviews*, Vol. 27, 2016, pp. 1–29. https://doi.org/10.1016/J.JPHOTOCHEMREV.2016.05.001

[124] Paul, S., Rahman, M. A., Sharif, S. Bin, Kim, J. H., Siddiqui, S. E. T., and Hossain, M. A. M., "TiO2 as an Anode of High-Performance Lithium-Ion Batteries: A Comprehensive Review towards Practical Application," *Nanomaterials 2022, Vol. 12, Page 2034*, Vol. 12, No. 12, 2022, p. 2034. https://doi.org/10.3390/NANO12122034

[125] Shi, H., Shi, C., Jia, Z., Zhang, L., Wang, H., and Chen, J., "Titanium Dioxide-Based Anode Materials for Lithium-Ion Batteries: Structure and Synthesis," *RSC Advances*, Vol. 12, No. 52, 2022, pp. 33641–33652. https://doi.org/10.1039/D2RA05442F

[126] Ren, L., Liu, Y., Qi, X., Hui, K. S., Hui, K. N., Huang, Z., Li, J., Huang, K., and Zhong, J., "An Architectured TiO2 Nanosheet with Discrete Integrated Nanocrystalline Subunits and Its Application in Lithium Batteries," *Journal of Materials Chemistry*, Vol. 22, No. 40, 2012, pp. 21513–21518. https://doi.org/10.1039/C2JM33085G

[127] Li, J., Huang, J., Li, J., Cao, L., Qi, H., and Cheng, Y., "N-Doped TiO2/RGO Hybrids as Superior Li-Ion Battery Anodes with Enhanced Li-Ions Storage Capacity," *Journal of Alloys and Compounds*, Vol. 784, 2019, pp. 165–172. https://doi.org/10.1016/J.JALLCOM.2019.01.061

[128] Han, Z., Peng, J., Liu, L., Wang, G., Yu, F., and Guo, X., "Few-Layer TiO2–B Nanosheets with N-Doped Graphene Nanosheets as a Highly Robust Anode for Lithium-Ion Batteries," *RSC Advances*, Vol. 7, No.

13, 2017, pp. 7864–7869. https://doi.org/10.1039/C6RA26929J

[129] Zhang, M., Yin, K., Hood, Z. D., Bi, Z., Bridges, C. A., Dai, S., Meng, Y. S., Paranthaman, M. P., and Chi, M., "In Situ TEM Observation of the Electrochemical Lithiation of N-Doped Anatase TiO2 Nanotubes as Anodes for Lithium-Ion Batteries," *Journal of Materials Chemistry A*, Vol. 5, No. 39, 2017, pp. 20651–20657. https://doi.org/10.1039/C7TA05877B

[130] Choi, W. H., Lee, C. H., Kim, H. eun, Lee, S. U., and Bang, J. H., "Designing a High-Performance Nitrogen-Doped Titanium Dioxide Anode Material for Lithium-Ion Batteries by Unravelling the Nitrogen Doping Effect," *Nano Energy*, Vol. 74, 2020, p. 104829. https://doi.org/10.1016/J.NANOEN.2020.104829

[131] Choi, C., Yang, S., and Kim, M., "TiO2/N-Doped Carbon Coating Derived from Recycled Waste Masks to Enhance the Electrochemical Performance of Silicon Anodes for Lithium-Ion Batteries," *Electrochimica Acta*, 2025, p. 146554. https://doi.org/10.1016/J.ELECTACTA.2025.146554

[132] Li, Y., Du, J., Sun, X., Lan, D., Cui, J., Zhao, H., Zhang, Y., and He, W., "Preparation of TiO2/Nitrogen-Doped CNF Composites as High-Performance Lithium-Ion Battery Anodes by Electrospinning," *Journal of Crystal Growth*, Vol. 624, 2023, p. 127417. https://doi.org/10.1016/J.JCRYSGRO.2023.127417

[133] Munonde, T. S., and Raphulu, M. C., "Review on Titanium Dioxide Nanostructured Electrode Materials for High-Performance Lithium Batteries," *Journal of Energy Storage*, Vol. 78, 2024, p. 110064. https://doi.org/10.1016/J.EST.2023.110064

[134] Opra, D. P., Sinebryukhov, S. L., Modin, E. B., Sokolov, A. A., Podgorbunsky, A. B., Ziatdinov, A. M., Ustinov, A. Y., Mayorov, V. Y., and Gnedenkov, S. V., "Manganese, Fluorine, and Nitrogen Co-Doped Bronze Titanium Dioxide Nanotubes with Improved Lithium-Ion Storage Properties," *Batteries*, Vol. 9, No. 4, 2023, p. 229. https://doi.org/10.3390/BATTERIES9040229/S1

[135] Hosseini, M., Momeni, M. M., and Faraji, M., "Electro-Oxidation of Hydrazine on Gold Nanoparticles Supported on TiO2 Nanotube

Matrix as a New High Active Electrode," *Journal of Molecular Catalysis A: Chemical*, Vol. 335, Nos. 1–2, 2011, pp. 199–204. https://doi.org/10.1016/J.MOLCATA.2010.11.034

[136]   Hosseini, M. G., and Momeni, M. M., "Platinum Nanoparticle-Decorated TiO2 Nanotube Arrays as New Highly Active and Non-Poisoning Catalyst for Photo-Electrochemical Oxidation of Galactose," *Applied Catalysis A: General*, Vols. 427–428, 2012, pp. 35–42. https://doi.org/10.1016/J.APCATA.2012.03.027

[137]   Babu, V. J., Vempati, S., Uyar, T., and Ramakrishna, S., "Review of One-Dimensional and Two-Dimensional Nanostructured Materials for Hydrogen Generation," *Physical Chemistry Chemical Physics*, Vol. 17, No. 5, 2015, pp. 2960–2986. https://doi.org/10.1039/C4CP04245J

[138]   Ge, M., Li, Q., Cao, C., Huang, J., Li, S., Zhang, S., Chen, Z., Zhang, K., Al-Deyab, S. S., and Lai, Y., "One-Dimensional TiO2 Nanotube Photocatalysts for Solar Water Splitting," *Advanced Science*, Vol. 4, No. 1, 2017, p. 1600152. https://doi.org/10.1002/ADVS.201600152

[139]   Zheng, J., Bao, S., Zhang, X., Wu, H., Chen, R., and Jin, P., "Pd-MgNix Nanospheres/Black-TiO2 Porous Films with Highly Efficient Hydrogen Production by near-Complete Suppression of Surface Recombination," *Applied Catalysis B: Environmental*, Vol. 183, 2016, pp. 69–74. https://doi.org/10.1016/J.APCATB.2015.10.031

[140]   Guo, W., Shen, Y., Wu, L., Gao, Y., and Ma, T., "Effect of N Dopant Amount on the Performance of Dye-Sensitized Solar Cells Based on N-Doped TiO 2 Electrodes," *J. Phys. Chem. C*, Vol. 115, 2011, pp. 21494–21499. https://doi.org/10.1021/jp2057496

[141]   Wang, H., Li, H., Wang, J., Wu, J., Li, D., Liu, M., and Su, P., "Nitrogen-Doped TiO2 Nanoparticles Better TiO2 Nanotube Array Photo-Anodes for Dye Sensitized Solar Cells," *Electrochimica Acta*, Vol. 137, 2014, pp. 744–750. https://doi.org/10.1016/J.ELECTACTA.2014.05.112

[142]   Fazli, F. I. M., Ahmad, M. K., Soon, C. F., Nafarizal, N., Suriani, A. B., Mohamed, A., Mamat, M. H., Malek, M. F., Shimomura, M., and Murakami, K., "Dye-Sensitized Solar Cell Using Pure Anatase TiO2 Annealed at Different Temperatures," *Optik*, Vol. 140, 2017, pp. 1063–

1068. https://doi.org/10.1016/J.IJLEO.2017.04.027

[143] Hou, X. ;, Aitola, K. ;, and Lund, P. D., "TiO 2 Nanotubes for Dye-Sensitized Solar Cells-A Review." https://doi.org/10.1002/ese3.831

[144] Wang, J., Tapio, K., Habert, A., Sorgues, S., Colbeau-Justin, C., Ratier, B., Scarisoreanu, M., Toppari, J., Herlin-Boime, N., and Bouclé, J., "Influence of Nitrogen Doping on Device Operation for TiO2-Based Solid-State Dye-Sensitized Solar Cells: Photo-Physics from Materials to Devices," *Nanomaterials*, Vol. 6, No. 3, 2016, p. 35. https://doi.org/10.3390/NANO6030035

[145] Tran, V. A., Truong, T. T., Phan, T. A. P., Nguyen, T. N., Huynh, T. Van, Agresti, A., Pescetelli, S., Le, T. K., Di Carlo, A., Lund, T., Le, S. N., and Nguyen, P. T., "Application of Nitrogen-Doped TiO2 Nano-Tubes in Dye-Sensitized Solar Cells," *Applied Surface Science*, Vol. 399, 2017, pp. 515–522. https://doi.org/10.1016/J.APSUSC.2016.12.125

[146] "Nitrogen-Doped TiO2 Nanoparticles Better TiO2 Nanotube Array Photo-Anodes for Dye Sensitized Solar Cells | Request PDF." Retrieved 11 June 2025. https://www.researchgate.net/publication/272928305_Nitrogen-doped_TiO2_nanoparticles_better_TiO2_nanotube_array_photo-anodes_for_dye_sensitized_solar_cells

[147] Qin, W., Lu, S., Wu, X., and Wang, S., "Dye-Sensitized Solar Cell Based on N-Doped TiO2 Electrodes Prepared on Titanium," *International Journal of Electrochemical Science*, Vol. 8, No. 6, 2013, pp. 7984–7990. https://doi.org/10.1016/S1452-3981(23)12863-X

[148] Dubey, R. S., Jadkar, S. R., and Bhorde, A. B., "Synthesis and Characterization of Various Doped TiO2 Nanocrystals for Dye-Sensitized Solar Cells," *ACS Omega*, Vol. 6, No. 5, 2021, pp. 3470–3482. https://doi.org/10.1021/ACSOMEGA.0C01614/ASSET/IMAGES/LARGE/AO0C01614_0010.JPEG

[149] Chen, X., and Mao, S. S., "Titanium Dioxide Nanomaterials: Synthesis, Properties, Modifications and Applications," *Chemical Reviews*, Vol. 107, No. 7, 2007, pp. 2891–2959. https://doi.org/10.1021/CR0500535/ASSET/CR0500535.FP.PNG_V03

[150] Zanoni, M. V. B., Irikura, K., Perini, J. A. L., Bessegato, G. G., Sandoval, M. A., and Salazar, R., "Recent Achievements in Photoelectrocatalytic Degradation of Pesticides," *Current Opinion in Electrochemistry*, Vol. 35, 2022, p. 101020. https://doi.org/10.1016/J.COELEC.2022.101020

[151] An, Y., Lei, T., Jiang, W., and Pang, H., "Research Progress on Photocatalytic, Electrocatalytic and Photoelectrocatalytic Selective Oxidation of 5-Hydroxymethylfurfural," *Green Chemistry*, Vol. 26, No. 21, 2024, pp. 10739–10773. https://doi.org/10.1039/D4GC03597F

[152] Wu, S., and Hu, Y. H., "A Comprehensive Review on Catalysts for Electrocatalytic and Photoelectrocatalytic Degradation of Antibiotics," *Chemical Engineering Journal*, Vol. 409, 2021, p. 127739. https://doi.org/10.1016/J.CEJ.2020.127739

[153] Alulema-Pullupaxi, P., Espinoza-Montero, P. J., Sigcha-Pallo, C., Vargas, R., Fernández, L., Peralta-Hernández, J. M., and Paz, J. L., "Fundamentals and Applications of Photoelectrocatalysis as an Efficient Process to Remove Pollutants from Water: A Review," *Chemosphere*, Vol. 281, 2021, p. 130821. https://doi.org/10.1016/J.CHEMOSPHERE.2021.130821

[154] Salmerón, I., Sharma, P. K., Polo-López, M. I., Tolosana, A., McMichael, S., Oller, I., Byrne, J. A., and Fernández-Ibáñez, P., "Electrochemically Assisted Photocatalysis for the Simultaneous Degradation of Organic Micro-Contaminants and Inactivation of Microorganisms in Water," *Process Safety and Environmental Protection*, Vol. 147, 2021, pp. 488–496. https://doi.org/10.1016/J.PSEP.2020.09.060

[155] Wang, F., Ou, R., Yu, H., Lu, Y., Qu, J., Zhu, S., Zhang, L., and Huo, M., "Photoelectrocatalytic PNP Removal Using C3N4 Nanosheets/α-Fe2O3 Nanoarrays Photoanode: Performance, Mechanism and Degradation Pathways," *Applied Surface Science*, Vol. 565, 2021. https://doi.org/10.1016/J.APSUSC.2021.150597

[156] Brito, J. F. de, Bessegato, G. G., Perini, J. A. L., Torquato, L. D. de M., and Zanoni, M. V. B., "Advances in Photoelectroreduction of CO2 to Hydrocarbons Fuels: Contributions of Functional Materials," *Journal of CO2 Utilization*, Vol. 55, 2022, p. 101810. https://doi.org/10.1016/J.JCOU.2021.101810

[157] Bastug Azer, B., Gulsaran, A., Pennings, J. R., Saritas, R., Kocer, S., Bennett, J. L., Devdas Abhang, Y., Pope, M. A., Abdel-Rahman, E., and Yavuz, M., "A Review: TiO2 Based Photoelectrocatalytic Chemical Oxygen Demand Sensors and Their Usage in Industrial Applications," *Journal of Electroanalytical Chemistry*, Vol. 918, 2022. https://doi.org/10.1016/J.JELECHEM.2022.116466

[158] Monllor-Satoca, D., Díez-García, M. I., Lana-Villarreal, T., and Gómez, R., "Photoelectrocatalytic Production of Solar Fuels with Semiconductor Oxides: Materials, Activity and Modeling," *Chemical Communications*, Vol. 56, No. 82, 2020, pp. 12272–12289. https://doi.org/10.1039/D0CC04387G

[159] Carneiro, P. A., Osugi, M. E., Sene, J. J., Anderson, M. A., and Zanoni, M. V. B., "Evaluation of Color Removal and Degradation of a Reactive Textile Azo Dye on Nanoporous TiO2 Thin-Film Electrodes," *Electrochimica Acta*, Vol. 49, Nos. 22-23 SPEC. ISS., 2004, pp. 3807–3820. https://doi.org/10.1016/J.ELECTACTA.2003.12.057

[160] Carneiro, P. A., Osugi, M. E., Fugivara, C. S., Boralle, N., Furlan, M., and Zanoni, M. V. B., "Evaluation of Different Electrochemical Methods on the Oxidation and Degradation of Reactive Blue 4 in Aqueous Solution," *Chemosphere*, Vol. 59, No. 3, 2005, pp. 431–439. https://doi.org/10.1016/J.CHEMOSPHERE.2004.10.043

[161] Zainal, Z., Lee, C. Y., Hussein, M. Z., Kassim, A., and Yusof, N. A., "Electrochemical-Assisted Photodegradation of Dye on TiO2 Thin Films: Investigation on the Effect of Operational Parameters," *Journal of Hazardous Materials*, Vol. 118, Nos. 1–3, 2005, pp. 197–203. https://doi.org/10.1016/J.JHAZMAT.2004.11.009

[162] Catanho, M., Malpass, G. R. P., and Motheo, A. J., "Photoelectrochemical Treatment of the Dye Reactive Red 198 Using DSA® Electrodes," *Applied Catalysis B: Environmental*, Vol. 62, Nos. 3–4, 2006, pp. 193–200. https://doi.org/10.1016/J.APCATB.2005.07.011

[163] Socha, A., Sochocka, E., Podsiadły, R., and Sokołowska, J., "Electrochemical and Photoelectrochemical Treatment of C.I. Acid Violet 1," *Dyes and Pigments*, Vol. 73, No. 3, 2007, pp. 390–393.

https://doi.org/10.1016/J.DYEPIG.2006.01.007

[164] Socha, A., Chrzescijanska, E., and Kusmierek, E., "Electrochemical and Photoelectrochemical Treatment of 1-Aminonaphthalene-3,6-Disulphonic Acid," *Dyes and Pigments*, Vol. 67, No. 1, 2005, pp. 71–75. https://doi.org/10.1016/J.DYEPIG.2004.10.012

[165] Socha, A., Sochocka, E., Podsiadły, R., and Sokołowska, J., "Electrochemical and Photoelectrochemical Degradation of Direct Dyes," *Coloration Technology*, Vol. 122, No. 4, 2006, pp. 207–212. https://doi.org/10.1111/J.1478-4408.2006.00027.X

[166] De Moura, D. C., De Araújo, C. K. C., Zanta, C. L. P. S., Salazar, R., and Martínez-Huitle, C. A., "Active Chlorine Species Electrogenerated on Ti/Ru0.3Ti0.7O2surface: Electrochemical Behavior, Concentration Determination and Their Application," *Journal of Electroanalytical Chemistry*, Vol. 731, 2014, pp. 145–152. https://doi.org/10.1016/J.JELECHEM.2014.08.008

[167] Janáky, C., Kecsenovity, E., and Rajeshwar, K., "Electrodeposition of Inorganic Oxide/Nanocarbon Composites: Opportunities and Challenges," *ChemElectroChem*, Vol. 3, No. 2, 2016, pp. 181–192. https://doi.org/10.1002/CELC.201500460

[168] Fernández, J., Kiwi, J., Baeza, J., Freer, J., Lizama, C., and Mansilla, H. D., "Orange II Photocatalysis on Immobilised TiO2: Effect of the PH and H2O2," *Applied Catalysis B: Environmental*, Vol. 48, No. 3, 2004, pp. 205–211. https://doi.org/10.1016/J.APCATB.2003.10.014

[169] Ku, Y., Lee, Y. C., and Wang, W. Y., "Photocatalytic Decomposition of 2-Chlorophenol in Aqueous Solution by UV/TiO2 Process with Applied External Bias Voltage," *Journal of hazardous materials*, Vol. 138, No. 2, 2006, pp. 350–356. https://doi.org/10.1016/J.JHAZMAT.2006.05.057

[170] Ye, S., Chen, Y., Yao, X., and Zhang, J., "Simultaneous Removal of Organic Pollutants and Heavy Metals in Wastewater by Photoelectrocatalysis: A Review," *Chemosphere*, Vol. 273, 2021, p. 128503. https://doi.org/10.1016/J.CHEMOSPHERE.2020.128503

[171] Bessegato, G. G., Guaraldo, T. T., de Brito, J. F., Brugnera, M. F., and

Zanoni, M. V. B., "Achievements and Trends in Photoelectrocatalysis: From Environmental to Energy Applications," *Electrocatalysis 2015 6:5*, Vol. 6, No. 5, 2015, pp. 415–441. https://doi.org/10.1007/S12678-015-0259-9

[172] Daghrir, R., Drogui, P., and Robert, D., "Photoelectrocatalytic Technologies for Environmental Applications," *Journal of Photochemistry and Photobiology A: Chemistry*, Vol. 238, 2012, pp. 41–52. https://doi.org/10.1016/J.JPHOTOCHEM.2012.04.009

[173] Wen, Y., Zhao, Y., Guo, M., and Xu, Y., "Synergetic Effect of Fe2O3 and BiVO4 as Photocatalyst Nanocomposites for Improved Photo-Fenton Catalytic Activity," *Journal of Materials Science*, Vol. 54, No. 11, 2019, pp. 8236–8246. https://doi.org/10.1007/S10853-019-03511-X

[174] Xu, Y., Jia, J., Zhong, D., and Wang, Y., "Degradation of Dye Wastewater in a Thin-Film Photoelectrocatalytic (PEC) Reactor with Slant-Placed TiO2/Ti Anode," *Chemical Engineering Journal*, Vol. 150, Nos. 2–3, 2009, pp. 302–307. https://doi.org/10.1016/J.CEJ.2009.01.002

[175] Zarei, E., and Ojani, R., "Fundamentals and Some Applications of Photoelectrocatalysis and Effective Factors on Its Efficiency: A Review," *Journal of Solid State Electrochemistry 2016 21:2*, Vol. 21, No. 2, 2016, pp. 305–336. https://doi.org/10.1007/S10008-016-3385-2

[176] Garcia-Segura, S., and Brillas, E., "Applied Photoelectrocatalysis on the Degradation of Organic Pollutants in Wastewaters," *Journal of Photochemistry and Photobiology C: Photochemistry Reviews*, Vol. 31, 2017, pp. 1–35. https://doi.org/10.1016/J.JPHOTOCHEMREV.2017.01.005

[177] Daghrir, R., Drogui, P., and El Khakani, M. A., "Photoelectrocatalytic Oxidation of Chlortetracycline Using Ti/TiO2 Photo-Anode with Simultaneous H2O2 Production," *Electrochimica Acta*, Vol. 87, 2013, pp. 18–31. https://doi.org/10.1016/J.ELECTACTA.2012.09.020

[178] Wang, J. L., and Xu, L. J., "Advanced Oxidation Processes for Wastewater Treatment: Formation of Hydroxyl Radical and Application," *Critical Reviews in Environmental Science and Technology*, Vol. 42, No. 3, 2012, pp. 251–325. https://doi.org/10.1080/10643389.2010.507698

[179] Carver, C., Ulissi, Z., Ong, C. K., Dennison, S., Kelsall, G. H., and Hellgardt, K., "Modelling and Development of Photoelectrochemical Reactor for H2 Production," *International Journal of Hydrogen Energy*, Vol. 37, No. 3, 2012, pp. 2911–2923. https://doi.org/10.1016/J.IJHYDENE.2011.07.012

[180] Lianos, P., "Review of Recent Trends in Photoelectrocatalytic Conversion of Solar Energy to Electricity and Hydrogen," *Applied Catalysis B: Environmental*, Vol. 210, 2017, pp. 235–254. https://doi.org/10.1016/J.APCATB.2017.03.067

[181] Orimolade, B. O., Koiki, B. A., Peleyeju, G. M., and Arotiba, O. A., "Visible Light Driven Photoelectrocatalysis on a FTO/BiVO 4 /BiOI Anode for Water Treatment Involving Emerging Pharmaceutical Pollutants," *Electrochimica Acta*, Vol. 307, 2019, pp. 285–292. https://doi.org/10.1016/J.ELECTACTA.2019.03.217

[182] Paschoal, F. M. M., Anderson, M. A., and Zanoni, M. V. B., "The Photoelectrocatalytic Oxidative Treatment of Textile Wastewater Containing Disperse Dyes," *Desalination*, Vol. 249, No. 3, 2009, pp. 1350–1355. https://doi.org/10.1016/J.DESAL.2009.06.024

[183] Selcuk, H., and Bekbolet, M., "Photocatalytic and Photoelectrocatalytic Humic Acid Removal and Selectivity of TiO2 Coated Photoanode," *Chemosphere*, Vol. 73, No. 5, 2008, pp. 854–858. https://doi.org/10.1016/j.chemosphere.2008.05.069

[184] Xu, Y. L., Zhong, D. J., Jia, J. P., Chen, S., and Li, K., "Enhanced Dye Wastewater Degradation Efficiency Using a Flowing Aqueous Film Photoelectrocatalytic Reactor," *Journal of Environmental Science and Health - Part A Toxic/Hazardous Substances and Environmental Engineering*, Vol. 43, No. 10, 2008, pp. 1215–1222. https://doi.org/10.1080/10934520802171790

[185] Mazierski, P., Borzyszkowska, A. F., Wilczewska, P., Białk-Bielińska, A., Zaleska-Medynska, A., Siedlecka, E. M., and Pieczyńska, A., "Removal of 5-Fluorouracil by Solar-Driven Photoelectrocatalytic Oxidation Using Ti/TiO2(NT) Photoelectrodes," *Water Research*, Vol. 157, 2019, pp. 610–620. https://doi.org/10.1016/J.WATRES.2019.04.010

[186] Zhao, X., Qu, J., Liu, H., Wang, C., Xiao, S., Liu, R., Liu, P., Lan, H.,

and Hu, C., "Photoelectrochemical Treatment of Landfill Leachate in a Continuous Flow Reactor," *Bioresource technology*, Vol. 101, No. 3, 2010, pp. 865–869. https://doi.org/10.1016/J.BIORTECH.2009.08.098

[187]   Cardoso, J. C., Bessegato, G. G., and Boldrin Zanoni, M. V., "Efficiency Comparison of Ozonation, Photolysis, Photocatalysis and Photoelectrocatalysis Methods in Real Textile Wastewater Decolorization," *Water research*, Vol. 98, 2016, pp. 39–46. https://doi.org/10.1016/J.WATRES.2016.04.004

[188]   de Brito, J. F., Bessegato, G. G., de Toledo e Souza, P. R. F., Viana, T. S., de Oliveira, D. P., Martínez-Huitle, C. A., and Zanoni, M. V. B., "Combination of Photoelectrocatalysis and Ozonation as a Good Strategy for Organics Oxidation and Decreased Toxicity in Oil-Produced Water," *Journal of The Electrochemical Society*, Vol. 166, No. 5, 2019, pp. H3231–H3238. https://doi.org/10.1149/2.0331905JES/XML

[189]   Zhou, X., Zheng, Y., Zhou, J., and Zhou, S., "Degradation Kinetics of Photoelectrocatalysis on Landfill Leachate Using Codoped TiO2/Ti Photoelectrodes," *Journal of Nanomaterials*, Vol. 2015, No. 1, 2015, p. 810579. https://doi.org/10.1155/2015/810579

[190]   "(PDF) Photo-Electrochemical Technologies for Removing Organic Compounds in Wastewaters." Retrieved 12 April 2025. https://www.researchgate.net/publication/325631557_Photo-Electrochemical_technologies_for_removing_organic_compounds_in_wastewaters

[191]   Xiao, K., Liang, H., Chen, S., Yang, B., Zhang, J., and Li, J., "Enhanced Photoelectrocatalytic Degradation of Bisphenol A and Simultaneous Production of Hydrogen Peroxide in Saline Wastewater Treatment," *Chemosphere*, Vol. 222, 2019, pp. 141–148. https://doi.org/10.1016/J.CHEMOSPHERE.2019.01.109

[192]   Cheng, L., Jiang, T., Yan, K., Gong, J., and Zhang, J., "A Dual-Cathode Photoelectrocatalysis-Electroenzymatic Catalysis System by Coupling BiVO4 Photoanode with Hemin/Cu and Carbon Cloth Cathodes for Degradation of Tetracycline," *Electrochimica Acta*, Vol. 298, 2019, pp. 561–569. https://doi.org/10.1016/J.ELECTACTA.2018.12.086

[193] Chen, Y., Zhao, X., Guan, W., Cao, D., Guo, T., Zhang, X., and Wang, Y., "Photoelectrocatalytic Oxidation of Metal-EDTA and Recovery of Metals by Electrodeposition with a Rotating Cathode," *Chemical Engineering Journal*, Vol. 324, 2017, pp. 74–82. https://doi.org/10.1016/J.CEJ.2017.05.031

[194] Butterfield, I. M., Christensen, P. A., Hamnett, A., Shaw, K. E., Walker, G. M., Walker, S. A., and Howarth, C. R., "Applied Studies on Immobilized Titanium Dioxide Films as Catalysts for the Photoelectrochemical Detoxification of Water," *Journal of Applied Electrochemistry*, Vol. 27, No. 4, 1997, pp. 385–395. https://doi.org/10.1023/A:1018453402332/METRICS

[195] Sclafani, A., Palmisano, L., and Schiavello, M., "Influence of the Preparation Methods of TiO2 on the Photocatalytic Degradation of Phenol in Aqueous Dispersion," *Journal of Physical Chemistry*, Vol. 94, No. 2, 1990, pp. 829–832. https://doi.org/10.1021/J100365A058/ASSET/J100365A058.FP.PNG_V03

[196] Xiang, C., Weber, A. Z., Ardo, S., Berger, A., Chen, Y. K., Coridan, R., Fountaine, K. T., Haussener, S., Hu, S., Liu, R., Lewis, N. S., Modestino, M. A., Shaner, M. M., Singh, M. R., Stevens, J. C., Sun, K., and Walczak, K., "Modeling, Simulation, and Implementation of Solar-Driven Water-Splitting Devices," *Angewandte Chemie International Edition*, Vol. 55, No. 42, 2016, pp. 12974–12988. https://doi.org/10.1002/ANIE.201510463

[197] Fang, T., Liao, L., Xu, X., Peng, J., and Jing, Y., "Removal of COD and Colour in Real Pharmaceutical Wastewater by Photoelectrocatalytic Oxidation Method," *Environmental technology*, Vol. 34, Nos. 5–8, 2013, pp. 779–786. https://doi.org/10.1080/09593330.2012.715760

[198] Li, G., An, T., Chen, J., Sheng, G., Fu, J., Chen, F., Zhang, S., and Zhao, H., "Photoelectrocatalytic Decontamination of Oilfield Produced Wastewater Containing Refractory Organic Pollutants in the Presence of High Concentration of Chloride Ions," *Journal of hazardous materials*, Vol. 138, No. 2, 2006, pp. 392–400. https://doi.org/10.1016/J.JHAZMAT.2006.05.083

[199] Wang, N., Li, X., Wang, Y., Quan, X., and Chen, G., "Evaluation of Bias Potential Enhanced Photocatalytic Degradation of 4-Chlorophenol with TiO2 Nanotube Fabricated by Anodic Oxidation Method," *Chemical Engineering Journal*, Vol. 146, No. 1, 2009, pp. 30–35. https://doi.org/10.1016/J.CEJ.2008.05.025

[200] Vargas, R., and Núñez, O., "Hydrogen Bond Interactions at the TiO2 Surface: Their Contribution to the PH Dependent Photo-Catalytic Degradation of p-Nitrophenol," *Journal of Molecular Catalysis A: Chemical*, Vol. 300, Nos. 1–2, 2009, pp. 65–71. https://doi.org/10.1016/J.MOLCATA.2008.10.029

[201] Martínez-Huitle, C. A., Rodrigo, M. A., Sirés, I., and Scialdone, O., "Single and Coupled Electrochemical Processes and Reactors for the Abatement of Organic Water Pollutants: A Critical Review," *Chemical reviews*, Vol. 115, No. 24, 2015, pp. 13362–13407. https://doi.org/10.1021/ACS.CHEMREV.5B00361

[202] Li, X. Z., Liu, H. L., Yue, P. T., and Sun, Y. P., "Photoelectrocatalytic Oxidation of Rose Bengal in Aqueous Solution Using a Ti/TiO2 Mesh Electrode," *Environmental Science and Technology*, Vol. 34, No. 20, 2000, pp. 4401–4406. https://doi.org/10.1021/ES000939K

[203] Yang, J., Chen, C., Ji, H., Ma, W., and Zhao, J., "Mechanism of TiO2-Assisted Photocatalytic Degradation of Dyes under Visible Irradiation: Photoelectrocatalytic Study by TiO2-Film Electrodes," *Journal of Physical Chemistry B*, Vol. 109, No. 46, 2005, pp. 21900–21907. https://doi.org/10.1021/JP0540914/SUPPL_FILE/JP0540914SI200 50908_052006.PDF

[204] An, T., Xiong, Y., Li, G., Zha, C., and Zhu, X., "Synergetic Effect in Degradation of Formic Acid Using a New Photoelectrochemical Reactor," *Journal of Photochemistry and Photobiology A: Chemistry*, Vol. 152, Nos. 1–3, 2002, pp. 155–165. https://doi.org/10.1016/S1010-6030(02)00211-3

[205] Zhou, M., and Lei, L., "An Improved UV/Fe3+ Process by Combination with Electrocatalysis for p-Nitrophenol Degradation," *Chemosphere*, Vol. 63, No. 6, 2006, pp. 1032–1040. https://doi.org/10.1016/J.CHEMOSPHERE.2005.08.057

[206] Di Paola, A., Augugliaro, V., Palmisano, L., Pantaleo, G., and Savinov, E., "Heterogeneous Photocatalytic Degradation of Nitrophenols," *Journal of Photochemistry and Photobiology A: Chemistry*, Vol. 155, Nos. 1–3, 2003, pp. 207–214. https://doi.org/10.1016/S1010-6030(02)00390-8

[207] Huang, J. Y., Zhang, K. Q., and Lai, Y. K., "Fabrication, Modification, and Emerging Applications of TiO2 Nanotube Arrays by Electrochemical Synthesis: A Review," *International Journal of Photoenergy*, Vol. 2013, No. 1, 2013, p. 761971. https://doi.org/10.1155/2013/761971

[208] Ge, M., Cao, C., Huang, J., Li, S., Chen, Z., Zhang, K. Q., Al-Deyab, S. S., and Lai, Y., "A Review of One-Dimensional TiO2 Nanostructured Materials for Environmental and Energy Applications," *Journal of Materials Chemistry A*, Vol. 4, No. 18, 2016, pp. 6772–6801. https://doi.org/10.1039/C5TA09323F

[209] Ge, M. Z., Cao, C. Y., Huang, J. Y., Li, S. H., Zhang, S. N., Deng, S., Li, Q. S., Zhang, K. Q., and Lai, Y. K., "Synthesis, Modification, and Photo/Photoelectrocatalytic Degradation Applications of TiO2 Nanotube Arrays: A Review," *Nanotechnology Reviews*, Vol. 5, No. 1, 2016, pp. 75–112. https://doi.org/10.1515/NTREV-2015-0049

[210] Quan, X., Ruan, X., Zhao, H., Chen, S., and Zhao, Y., "Photoelectrocatalytic Degradation of Pentachlorophenol in Aqueous Solution Using a TiO2 Nanotube Film Electrode," *Environmental Pollution*, Vol. 147, No. 2, 2007, pp. 409–414. https://doi.org/10.1016/J.ENVPOL.2006.05.023

[211] Hohenberg, P., and Kohn, W., "Inhomogeneous Electron Gas," *Physical Review*, Vol. 136, No. 3B, 1964, p. B864. https://doi.org/10.1103/PhysRev.136.B864

[212] Kohn, W., and Sham, L. J., "Self-Consistent Equations Including Exchange and Correlation Effects," *Physical Review*, Vol. 140, No. 4A, 1965, p. A1133. https://doi.org/10.1103/PHYSREV.140.A1133/FIGURE/1/THUMB

[213] "Density-Functional Theory of Atoms and Molecules - Robert G. Parr; Yang Weitao - Oxford University Press." Retrieved 7 July 2025.

https://global.oup.com/academic/product/density-functional-theory-of-atoms-and-molecules-9780195092769?cc=us&lang=en&

[214] Jones, R. O., and Gunnarsson, O., "The Density Functional Formalism, Its Applications and Prospects," *Reviews of Modern Physics*, Vol. 61, No. 3, 1989, p. 689. https://doi.org/10.1103/RevModPhys.61.689

[215] Jochym, D., and Refson, K., "Lattice Dynamics and Spectroscopy from DFT," No. May 2012. Retrieved 20 September 2025. https://books.google.com/books/about/Electronic_Structure.html?id=dmRTFLpSGNsC

[216] Burke, K., "Perspective on Density Functional Theory," *The Journal of Chemical Physics*, Vol. 136, No. 15, 2012. https://doi.org/10.1063/1.4704546

[217] Becke, A. D., "Perspective: Fifty Years of Density-Functional Theory in Chemical Physics," *The Journal of Chemical Physics*, Vol. 140, No. 18, 2014, pp. 18–301. https://doi.org/10.1063/1.4869598

[218] Sevak, R., "Engineering the Electronic Structure in Titanium Dioxide via Scandium Doping Based on the Density Functional Theory Approach for the Photocatalysis and Photovoltaic Applications."

[219] Janotti, A., Varley, J. B., Rinke, P., Umezawa, N., Kresse, G., and Van De Walle, C. G., "Hybrid Functional Studies of the Oxygen Vacancy in <math Xmlns="http://Www.W3.Org/1998/Math/MathML" Display="inline"><mrow><msub><mrow><mtext>TiO</Mtext></Mrow><mn>2</Mn></Msub></Mrow></Math>," *Physical Review B*, Vol. 81, No. 8, 2010, p. 085212. https://doi.org/10.1103/PhysRevB.81.085212

[220] Jochym, D., and Refson, K., "Lattice Dynamics and Spectroscopy from DFT," No. May 2012. Retrieved 12 April 2025. https://books.google.com/books/about/Electronic_Structure.html?id=dmRTFLpSGNsC

[221] Dreizler, R. M. ., and Gross, E. K. U. ., "Density Functional Theory," 2012. Retrieved 12 April 2025. https://books.google.com/books/about/Density_Functional_Theory

.html?id=t6PvCAAAQBAJ

[222] Calais, J.-L., "Density-Functional Theory of Atoms and Molecules. R.G. Parr and W. Yang, Oxford University Press, New York, Oxford, 1989. IX + 333 Pp. Price £45.00," *International Journal of Quantum Chemistry*, Vol. 47, No. 1, 1993, pp. 101–101. https://doi.org/10.1002/QUA.560470107

[223] Perdew, J. P., Burke, K., and Ernzerhof, M., "Generalized Gradient Approximation Made Simple," *Physical Review Letters*, Vol. 77, No. 18, 1996, p. 3865. https://doi.org/10.1103/PhysRevLett.77.3865

ABOUT AUTHORS

Hakimeh Ghazaei is pursuing her Ph.D. in Materials Engineering at the University of Tabriz, where she investigates one-dimensional TiO_2 nanotube arrays, nitrogen/defect-engineered photocatalysts, and their integration in photoelectrocatalysis, dye-sensitized solar cells, and supercapacitor anodes. Her expertise includes anodic fabrication, post-treatment/doping strategies, and structure–property correlations supported by XRD, Raman, SEM/EDS, UV–Vis, EIS, and PEC measurements. She has co-authored works on N-doped TiO_2 electrodes and black titania, with an emphasis on translating lab-scale insights into scalable processes.

Dr. Shahab Khameneh Asl is an Iranian materials scientist and researcher at the University of Tabriz, where he has served since 2016 as Head of the Ceramic Laboratory Complex in the Department of Materials Engineering. He earned his PhD from the Materials and Energy Research Center at the age of 28 and previously studied new materials at Sharif University of Technology. His research spans nanostructured ceramics, photocatalysts, electroceramics, energy materials, surface engineering, and corrosion and tribological systems, with published work on nitrogen-doped TiO_2 nanotubes, graphene/MoS_2 composites for flexible energy storage, and photocatalytic water treatment. Overseeing multiple specialized labs and mentoring a large team of faculty, doctoral candidates, and students, Dr. Khameneh Asl has been recognized internationally for his contributions, including receiving the Green Energy Breakthrough Research Award for his innovative work in climate action and sustainable energy.

Dr. Shahin Khameneh Asl is an Associate Professor in the Department of Materials Engineering at the University of Tabriz, Iran. His research focuses on materials science with expertise in coatings, corrosion, surface engineering, and nanostructured materials, and he has published more than 50 scientific articles. He has contributed to studies on TiO$_2$ nanotubes, nitrogen-doped semiconductors, high-entropy alloy thin films, and natural protein-based corrosion inhibitors, establishing a strong profile in advanced material characterization and surface technologies.

The editorial oversight for this book was provided by **Dr. Saeid Kakooei** (Editor). Dr. Saeid Kakooei is an accomplished materials scientist and corrosion engineer with over 20 years of experience spanning academia, industry, and research leadership. He holds a PhD in Mechanical Engineering (Advanced Materials & Corrosion) from Universiti Teknologi Petronas, Malaysia, and has built a career at the intersection of advanced materials, corrosion, nanotechnology, and manufacturing. Currently a Paint Engineer II at Subaru of Indiana Automotive, he previously served as a Senior Research Engineer and Lab Manager at Purdue University's School of Materials Engineering, where he led multimillion-dollar DOE- and NSF-funded projects and co-founded Robus Ceramic Inc. to translate research into commercialization. Earlier, he was Head of the Centre for Corrosion Research at Universiti Teknologi Petronas, where he also taught as Senior Lecturer. Dr. Kakooei has published more than 65 peer-reviewed articles, co-edited several Elsevier volumes, holds patents including a pH sensor for corrosion monitoring, and has over 3,000 citations (h-index 27). He is an active member of professional bodies such as NACE (where he chaired the Technical Program Committee) and the Institute of Materials, Malaysia, and serves on editorial boards for multiple international journals.

www.ingramcontent.com/pod-product-compliance
Lightning Source LLC
Chambersburg PA
CBHW061259220326
41599CB00028B/5706